Critical Factors for Berth Productivity in Container Terminal

Bo Lu · Shouyang Wang

Critical Factors for Berth Productivity in Container Terminal

Science Press
Beijing

Springer

Bo Lu
School of Economics and Management
Dalian University
Shahekou District, Dalian
China

Shouyang Wang
School of Economics and Management
University of Chinese Academy of Sciences
Haidian District, Beijing
China

ISBN 978-981-10-2430-6 ISBN 978-981-10-2431-3 (eBook)
DOI 10.1007/978-981-10-2431-3

Jointly published with Science Press

Library of Congress Control Number: 2016948770

Printed on acid-free paper

This Springer imprint is published by Springer Nature
The registered company is Springer Nature Singapore Pte Ltd.
The registered company address is: 152 Beach Road, #22-06/08 Gateway East, Singapore 189721, Singapore

Contents

List of Figures

List of Tables

List of Formulations

Abstract

Globalization has led to the increasing importance of port performance and the awareness of the high-value industry for container terminals leads competitiveness of container terminals to keep high fiercely. In regard to competitive factors of container terminals, the most important point among several factors is seemed to be the speed of container loading and unloading on quayside. In container terminals in Asia, the berth productivity is an important concept in measuring performance and also shows big differences even though its condition is similar to each terminal.

The objective of this paper is to compare the Korea and China mega-container terminals to find the relationship and order of critical factors for berth productivity, which is dependent upon the capability, quantity of quay crane, transfer vehicle, and so on.

For this purpose, this research has researched related literatures and collected data about Korea and China major container terminals. Furthermore, this research tested sensitive analysis to evaluate the extent of berth productivity by changing independent variable. And then this research has established the regression model to evaluate which factor has had the biggest impact on productivity. The results of this study will give terminal operators guideline to improve berth productivity.

Chapter 1
Introduction

1.1 Background and Goal of Research

A recent competition between container ports and terminals, especially terminal productivity, is rapidly increasing. How to improve the productivity of container ports and terminals is, therefore, a challenge of port management. In order to enhance port competitiveness, fierce efforts for operation efficiency are being made as following directions, the infrastructure reinforcement through berth expansion, the improvement of stevedoring equipment and the increase of equipments, and the intensification of information technology and terminal automation. In a result, each terminal has invested lots of budgets to improve productivity. The examples for productivity improvement has been shown in different types, for example the lengthy arm of quay crane reaching 24 bays, high lifting capacity quay crane like twin-lift and tandem-lift, RFID technology for identifying vehicle and container in remote distance, dual cycle operation and pooling system of yard tractors for equipment productivity, the high capacity yard crane like RMGC and High Stacking System etc. can be listed.

In terms of terminal productivity, different capacity performance measures, such as berth length utilization rate (TEU/berth meter), total terminal area utilization rate (TEU/ha of total terminal area), quay crane (QC) utilization rate (TEU/QC), QC productivity (TEU/QC-hour) and average number of QCs per berth, may be used to compare the efficiency of the terminals in port. Generally speaking, there have been are two types of research trends in measuring productivity which is largely divided with the traditional productivity analytical method and data envelopment analysis. The traditional method is referred to area with the facility productivity analytical method and the synthetic method of productivity. The data envelopment analysis, DEA is commonly used to evaluate the efficiency of a number of producers. A typical statistical approach is characterized as a central tendency approach and it evaluates producers relative to an average producer. In contrast, DEA is an extreme point method and compares each producer with only the "best" producers. And

B. Lu and S. Wang, *Critical Factors for Berth Productivity in Container Terminal*,
DOI 10.1007/978-981-10-2431-3_1

Table 1.1 Top 50 container terminals in the world

Seq (05)	Port	Country	Throughput (05)	Throughput (04)	Increase ratio (%)
1	Singapore	Singapore	23,190,000	21,329,000	8.7
2	Hong Kong	China	22,602,000	21,930,000	3.1
3	Shanghai	China	18,084,000	14,554,000	24.3
4	Shenzhen	China	16,197,000	13,620,000	18.9
5	Busan	Korea	11,840,445	11,491,968	3.0
6	Kaohsiung	China	9,471,056	9,710,000	2.5
7	Rotterdam	Netherland	9,287,000	8,220,000	13.0
8	Hamburg	Germany	8,087,545	7,003,479	15.5
9	Dubai	UAE	7,619,219	6,428,884	18.5
10	Los Angeles	US	7,484,624	7,321,440	2.2
11	Long Beach	US	6,709,818	5,779,852	16.1
12	Antwerp	Belgium	6,482,029	6,063,746	6.9
13	Qingdao	China	6,307,000	5,139,000	22.7
14	Port Klang	Malaysia	5,543,527	5,243,593	5.7
15	Ningbo	China	5,208,000	4,000,000	30.2
16	Tianjin	China	4,801,000	3,810,000	26.0
17	New York	US	4,785,000	4,478,480	6.8
18	Guangzhou	China	4,683,000	3,313,200	41.3
19	Tanjung Pelepas	Malaysia	4,177,122	4,010,037	4.2
20	Laem Chabang	Thailand	3,793,802	3,621,595	4.8
21	Tokyo	Japan	3,759,000	3,580,096	5.0
22	Bremerhaven	Germany	3,700,000	3,469,104	6.7
23	Xiamen	China	3,342,300	2,871,700	16.4
24	Jakarta	Indonesia	3,281,580	3,170,000	3.5
25	Algeciras	Spain	3,179,614	2,937,381	8.2
26	Gioia tauro	Italy	3,160,981	3,261,034	3.1
27	Jeddah	Saudi Arabia	2,874,937	2,425,930	18.5
28	Yokohama	Japan	2,873,288	2,717,585	5.7
29	Felixstowe	UK	2,750,000	6,625,000	4.8
30	Panama	Panama	2,718,470	2,406,741	13.0
31	Dalian	China	2,690,000	2,210,000	21.7
32	Jawaharlal Nehru	India	2,670,000	2,370,000	12.7
33	Manila	Philippines	2,625,148	2,616,783	0.3
34	Salalah	Oman	2,492,000	2,228,541	11.8
35	Nagoya	Japan	2,491,194	2,303,539	8.1
36	Colombo	Sri Lanka	2,450,000	2,200,000	11.4
37	Valencia	Spain	2,410,000	2,145,000	12.4
38	Oakland	US	2,272,525	2,047,504	11.0

(continued)

Table 1.1 (continued)

Seq (05)	Port	Country	Throughput (05)	Throughput (04)	Increase ratio (%)
39	Santos	Brazil	2,267,921	1,882,639	20.5
40	Kobe	Japan	2,262,066	2,262,066	3.8
41	Sharjah	UAE	2,157,899	2,003,620	7.7
42	Le Havre	France	2,100,000	2,150,000	2.3
43	Osaka	Japan	2,094,277	2,009,150	4.2
44	Keelung	China	2,091,458	2,070,192	1.0
45	Seattle	US	2,088,000	1,790,000	16.6
46	Barcelona	Spain	2,071,480	1,916,493	8.1
47	Tacoma	US	2,066,447	1,800,000	14.8
48	Ho Chi Minh	Vietnam	2,030,508	1,685,770	20.4
49	Virginia	US	1,981,955	1,808,933	9.6
50	Charleston	US	1,980,000	1,863,916	6.2

Source The top 50 Anniversary of Container Shipping The Admiral Sheldon Kinney (USN) Lectures

which presents the combined analysis about the multiple input factors and multiple output factors, is classified in CCR models, which Charnes and Cooper and Rhodes etc. presented in 1978, and BCC models which Banker et al. presented in 1984. But the traditional method has difficulty in treating various variables synthetically. And DEA, which easily analyzes relative comparison between terminals, including the general factors (CCR model), yet has difficulty in examining the relationship of the each variable.

So, this study intended to find out relationship between productivity indicator and its factors impacting productivity, and then compares the difference between China and Korea mega-container terminal in order to give the guideline to terminal operators for the purpose of improving berth productivity.

In order to select research target in the paper, there is need to survey the throughput of Chinese and Korean container terminal. The top 50 container terminals in the world in Table 1.1.

1.2 Research Method and Composition

The factors that affect productivity of the container terminals are composed of infrastructure, superstructure, human resources and information structure, and automatic technologies.

In this study these considerable factors are discovered through the process of literature review, discussion with experts in working in container terminal for more than 20 years, international web site and brainstorming. And these factors are classified into few categories and then tried to collect related data in container terminals in South Korea and China. A regression analysis is executed based on collected data. Then this study intends to find the relationship between, dependent variables which are terminal productivity, and independent variables which affect them in China and Korea container terminals respectively, and then the paper will compare the difference between the two counties' analysis results.

This study considers the container terminal productivity with unit productivity. The unit productivity means gross berth productivity (GBP), which is defined with formula which divides annual total throughput of the terminal by total berthing time and number of berths in standard dimension. The applied data is gathered from 13 container terminals of Korea and 9 container terminals of China. In the process of data analysis, some data were found to be difficulty to measure, some data were found to be duplicated, some data were found to be insufficient significance, some data were found to be similar performance and they had been excluded from statistical analysis Fig. 1.1.

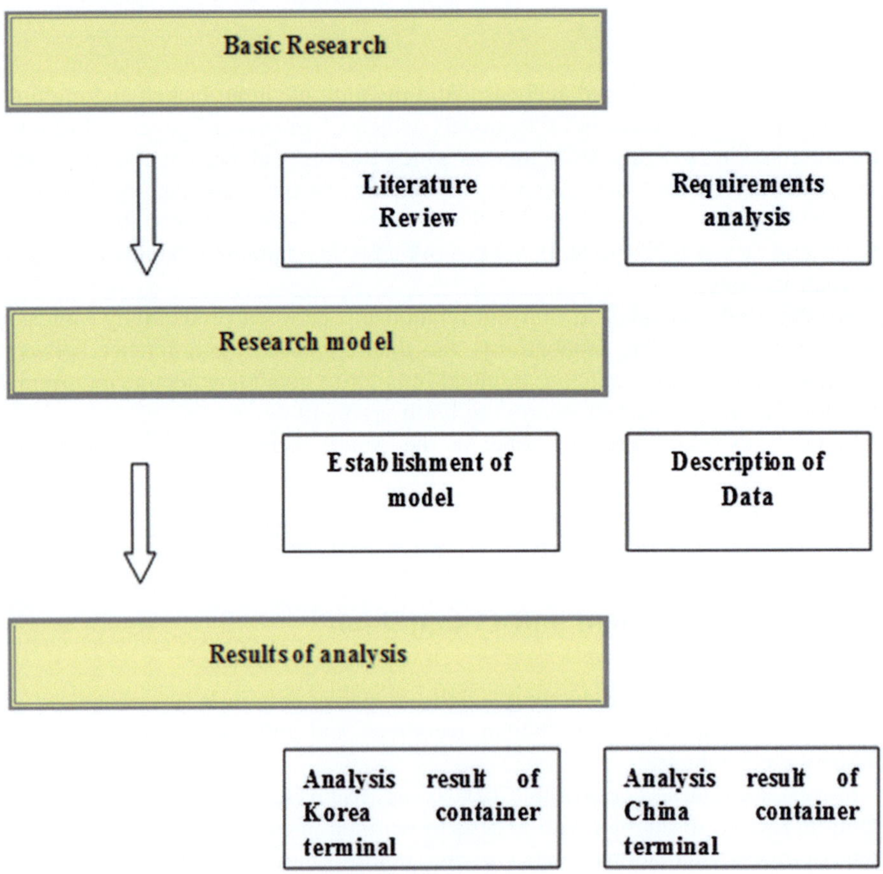

Fig. 1.1 Research framework

The composition of this study is as follows. First, Chap. 1 is introduction section: background, goal, composition and method of research. In Chap. 2, the collected literature will be reviewed in terms of the container terminal productivity. In Chap. 3, research model will be reviewed and the developmental status quo of container ports nowadays in the world also will be reviewed in a step of the research elaboration. The container terminals include the port of Shanghai, the port of Shenzhen, and the port of HongKong in China and the port of Busan, and the port of Gwangyang in Korea. In addition, the concept of container terminal productivity, the scope of modeling, modeling method, method of regression analysis, independent variables, and description of data will also be reviewed in this section. Chapter 4 contains the result of regression analysis based on the model and data which include the Analysis result of Port of Korea (Busan and Gwangyang) and China (Shanghai, Shenzhen, and Hong Kong), then I shall Compare the analysis result of port between China port and Korea port to get the comparative results. In Chap. 5, the result of research, its limitation and implication has been summarized.

Chapter 2
Literature Review

In order to find out critical factors for berth productivity in container terminals, we have collected and analyzed many related literatures. After we have found related papers which help to guide our understanding of concept of productivity, a few papers are selected for reference to define productivity factors.

First on the aspect of research method using model, we can find in many literatures that simulation model have been widely used. There are a few literatures using analytical modeling technique to analyze the factors that affect port productivity in container terminals. Nam Kyu Park et al. try to find the relationship and order of critical factors for berth productivity, which is dependent upon the capability, quantity of quay crane, transfer vehicle, and so on. Furthermore, he tested sensitive analysis to evaluate the extent of berth productivity by changing independent variable. And then he established the regression model to evaluate which factor has had the biggest impact on productivity.

Manuel Acosta et al. aim to analyze the factors that affect port competitiveness in case of the port of Algeciras Bay (PAB). It is essential to understand the perspective of the suppliers of port services, in order to apply the necessary judgment when formulating polices to strengthen the competitive position of a port. The model selected to meet the objective outlined is based on the 'extended diamond' of porter. The universe for the study comprises all the institutions and companies involved in the containerization process in the port of Algeciras Bay. First, within this methodology, two procedures are utilized from the quantitative perspective. A survey has been conducted to obtain responses to two groups of questions, one dealing with situation and the other with perceptions. Second, from a qualitative perspective, several open interviews have been conducted to corroborate the responses obtained to the questionnaire. From the data which are obtained in the survey, model has been estimated that minimizes the sum of the residuals in absolute value, utilizing the L1 regression; the model enables the competitive advantages and disadvantages of the PAB, from the point of view of the port operators, to be determined by an analysis of residuals. This analysis demonstrates that the variables representing greater competitive advantages are associated with

B. Lu and S. Wang, *Critical Factors for Berth Productivity in Container Terminal*,
DOI 10.1007/978-981-10-2431-3_2

transshipment, and with the maritime accessibility of the port to vessels, whereas those generating greater disadvantages are associated with rail transport which according to those surveyed, has resulted in the diversion of significant volumes of traffic to other Spanish or European ports with better communications, particularly Valencia, Barcelona and Gioia-Tauro, and with the state of competition in container trans-shipment activities. According to their research, we can see among the variables that infrastructure, superstructure, technology and communications systems, internal competition, and cooperation of the institutions and companies are the most contributed to the competitive advantage of the port. Nam-Kyu Park et al. aim at finding out the factors to enhance the productivity of container handling of quay crane, using simulation technique and statistical technique, three levels of decision making in terminal operation, strategy, and tactics and operation are selected for defining parameters of simulation. The result of the simulation and test shows that the significant factors to improve the productivity are the stack height of container, block dispersion and the distance in yard planning for shipment. Decision making in the operation level, however, is of significance in the mixed condition of strategic and tactical level. For this paper, ARENA, 4th version Kelton et al. is used to simulate. The reason for utilizing this simulation package is that it has some strong points for modeling the process such as ship arrival and queuing, container crane and yard tractor service pattern, etc. The scopes of the simulation include only container shipment handling on the yard and vessel. If the simulation starts, the simulated transfer crane works for container handling in the yard and waits until next job. As next job, the waiting yard truck receives the shipment and transports it to container crane. Last, container crane handles the shipment for loading on the vessel. To design the simulation model, the entity and its attributes are to be defined. As the entities for simulation, transfer crane, container crane and yard truck are chosen. Both of the processing distribution of container crane and transfer crane are normal distribution with least square error. The processing time of transfer crane and container crane and the velocity of yard truck will be used for simulation. Finally, the study finds out the main factors to improve problem. The results of the simulation and test shows that significant factors in productivity were the stacking height of container block dispersion and distance in yard planning for loading. As literatures using analytical modeling technique to improve the productivity and efficiency of container terminal, we will review Kim and colleague's researches. Kim et al. tried to solve the total travel time of straddle carrier to improve the productivity of container terminal. The authors tried to solve the routing problem by using integer programming technique. In addition, an efficient optimizing algorithm was also developed for solving routing problem of straddle carrier. Kim et al. suggested a decision container terminal support system to improve the efficient operation of port container terminal to enhance productivity.

Ceres Paragon Terminals claims that they are operating the fastest container terminal in the world. It will be the first marine terminal to integrate the systems provided by Navis with image acquisition portals, weigh-in-motion scales, closed-circuit television for security and gate control, remain-in-truck data entry by truck drivers, container/yard equipment positioning through satellite Global Positioning Systems (GPS), and advanced crane control systems.

Bin Liu explores China container port throughput correlated to the economic factors such as the GDP, foreign trade volume, fixed assets investment, interest rate and the exchange rate. The correlation analysis is conducted by the means of the double-log linear regression model with the help of the SPSS software based on the past 16 years data accumulated from the indexes. The research shows that China's container port throughput has a positive correlation with foreign trade volume and port fixed assets investment, negative correlation with GDP and interest rate. The result contradicts the international prevailing theory which is widely held among the port authorities.

Yun Cai et al. build a simulation and optimization model for minimizing the total stay time of ships, the paper builds a simulation and optimization model for minimizing the total stay time of ships, which aims at the problem of berth and quay crane scheduling. The simulation and optimization method creates and evaluates berth allocation schemes by genetic algorithms, obtains feasible solutions meeting berth restriction and quay crane scheduling strategy by simulation model, and achieves a good solution including berthing times, berthing positions and the number of quay cranes allocated to ships. The speediness and validity of the method is illustrated by an instance. By this way berth productivity can be enhanced.

Arie Sachish, has implemented in Israeli ports between 1966 and 1990, about cargo handling in ports. Research was undertaken with the goal of developing productivity functions that explain the changes in the productivity in Israeli ports by means of changes in various explanatory factors. A linear programming model is generated for developing productivity functions with an objective function of minimum deviations between calculated and actual productivity. The changes of volume, labour, capital, technology, management, and externalities had a great impact on productivity and on the effect of explanatory factors. The analysis of the relationship between the explanatory factors and actual productivity gives importance to the better understanding of organizational impact on productivity. This understanding can be used as a management tool through which a decision making process may be developed to arrive at decision on steps for raising productivity.

And on comparative analysis, we can find in many literatures that the authors take advantage of comparative analysis to research on port productivity. Peng Chuan-Sheng this paper through collecting china and foreign mega-container terminal statistical data, aims to analyze container terminal productivity. Firstly, the author define the concept of container terminal productivity, which has relationship between the port production output and input, and presents container terminal productivity depends on container terminal establishment, equipment, capacity of workers, work efficiency, and tactics of organization. Moreover, the author presents the indexes of scaling container terminal productivity which are efficiency of length of berth utilizing, efficiency of quay crane utilizing, and efficiency of container yard utilizing. At last, the author compares the indexes of China container terminal productivity with foreign mega-container terminal, and then indicates the developing potential and direction of China container terminal. Yang Chen, the paper in the context of concerning Asia and Europe with respect to container terminal operations, introduced the advanced production technology and its application. And author summarized some of the trends observed within this field. Because of China modern container terminals which have the limited resources e.g. area and facilities, the authors present that China modern container terminal should take advantage of the high-tech means, for instances, information technology, bigger-facilities, and automatic technology and facilities to enhance the container terminal productivity.

Besides, there are wide ranges of papers devoted to different aspects of productivity in container terminals. Mu Gu the author presents that nowadays because of productivity were not been improved and many carriers departed from the port which result in a sharply fall in throughput, for the propose of enhancing the port productivity and throughput, Singapore port takes several measures: making use of the advantaged geographic location, Singapore port enhances the management. For avoiding phenomenon of containers jam, Singapore port have expanded foundational establishment, added the deep water berth, enhanced the efficiency of loading and unloading, and also reduced turnaround time.

Carlos pestana and Nicolas peypoch the paper proposes a framework for benchmarking Italian and Portuguese seaports and the rationalization of their operational activities. The analysis is based on the Luenberger productivity indicator. A key advantage of this method is that it allows for both input contraction and output expansion in determining relative efficiencies of those seaports that perform worse than others. Several interesting and useful managerial insights and implications arise from the study. The general conclusion is that the benchmarking of multi-country seaports of a single country, reflecting different cultural by benchmarking the seaports of a single country, reflecting different cultural traditions and managerial practices. The paper aimed to extend the established literature on seaport productivity by applying the Luenberger indicator to estimate and decompose productivity change. Earlier studies on seaport productivity tend to employ non-parametric techniques and Malmquist productivity indexes. The Luenberger productivity indicator is a difference-based index of directional distance functions whereas the Malmquist index is a ratio-based measure. Luenberger introduces the shortage function-which has the desirable properties of accounting for both input contractions and output improvements-and establishes the duality between the shortage function and the profit function. Thus, the Luenberger indicator can accommodate either an input or output perspective corresponding to cost minimization or profit maximization. In the paper he employs the Luenberger productivity indicator of Chambers to estimate and decompose productivity growth on observation of Italian and Portuguese seaports between 2002 and 2004. A key advantage of this method is that it allows for both input contraction and output expansion in determining relative efficiencies and productivity changes. The general conclusion is that there is productivity growth in the majority of seaports analyzed, which is driven more by improvements in technological change than improvement in technical efficiency. Possible explanation for this feature of the results is that investment is not matched by upgraded managerial practices.

Prabir De presented and defined total factor productivity (TFP) model which is a measure of technological change in industry. His study assesses the total factor productivity (TFP) growth of the Indian port sector. TFP is a measure of technological change in industry. His study shows that ports in India are becoming more capital intensive in the post-liberalization period, and except one, the rest of all Indian ports came out with statistically significant TFP changes in last quarter century. His study also finds that, contrary to popular belief, the economy climate in post-reform period has yet to make any substantial impact on the performance of Indian ports in terms of the most popular measure of performance, TFP. Furthermore, the authors collected data from 1980–1981 to 2002–2003 in India port, following the Perpetual Inventory Accumulation (PIA) method. By researching the results indicate that there are tremendous spurt in international research on the relationship between globalization and productivity. Prabir De, presented port should put both skilled workers and application of higher technology in operation to improve the productivity and keep competitive.

Peng-fei Zhou, the author make an elaborate simulator was proposed about the logistics related to the arrival, berthing and departure of vessels in container terminal for designers and decision-makers to evaluate container terminal size in terms of berths and quay-cranes and its current performance. The simulator mainly includes simulation scenario generator, berth and quay-crane assignment simulation as well as input and output data planning. The simulation implement procedure consisting of several steps: first one should choose terminal size, horizon of simulation and throughput in the horizon, the basic berth number and crane number embody terminal size that is an important evaluation parameter. Generally, the horizon, which can determine the simulation scope, is set as a long period, such as several months, a year and even several years, in order to gain a satisfactory and accordant simulation result. The throughput is determined mainly according to the terminal size and the horizon. The second step is that generate traffic scenario once the algorithm for ship traffic scenario generation is in place, the form and parameters of the distributions used have to be determined. When adequate existing data are available, they are used for distribution fitting. Otherwise, projections of traffic flow, empirical distributions, and simplifying assumptions can compensate for the lack of detailed information. Container arrival distributions of year and month must be determined. And their measure unit is also chosen, which can result in errors of given distribution of each week, vessel arrival distribution of each week, one of each day and one of each time slot, and weather frequency each day by month are input. The third step is that allocate berths and quay-cranes and input terminal performance statistics data, the crane productivity show small differences from the terminals and the gross hourly productivity ranged from 22.82 to 24.61 moves. Apparently, the productivity have been affected by weather conditions. Therefore, the hourly productivity of 25 is used for the simulation. The last step is that observe the statistics data, in the statistics output, terminal key performance measure can be find on the basis of which the terminal is evaluated. If the performance measure cannot satisfy decision-makers or designers, we can return to step 1, and modify the input file. The simulator can be used to design additional terminal size, evaluate current terminal performance and analyze its operation cost.

Chien-Chang Chou, the objective of this paper is to analyze the container throughputs of major ports in Far Eastern region. These major ports include Kaohsiung port, Hong Kong port, Shanghai port, Pusan port, Kobe port and Singapore port. The findings lead to answer questions whether Taiwan province needs some new port policies, and how Kaohsiung port improves its competition as well. The import and export containers in China are increasing rapidly based on her strong economic growth in recent years. The container throughput of Hong Kong port, Shanghai port and Pusan port grew rapidly in recent years based on the large import and export container volume in South China and North China. For this reason, the market share of Hong Kong, Shanghai port increased obviously over the past 20 years. And Pusan port also kept almost the same market share. Thus the author suggests the Taiwan province shall improve the direct-transportation operation and port policies between Taiwan and mainland China. By the way, there will be more China's containers to go through Kaohsiung port for transshipment, just like Hong Kong port and Pusan port. On the other hand, although Kaohsiung port successively maintained the world's third largest container port before 2000, unfortunately, its market share in Far Eastern region decreased over past 20 years. For this reason, Taiwan province should try to reduce the transshipment cost of Kaohsiung port and attract more transshipment containers from Southeast countries. By the way, Kaohsiung port could improve its competition by maintaining higher container throughput, higher frequency of liners and lower transshipment cost.

"Critical Success Factors in Developing a Port and Hinterland as a Transshipment Hub", the author suggest that the developing a port should do these ways: (1) To develop a transshipment hub, geography a significant factor; (2) First class infrastructure; (3) Competitive pricing; (4) Create a business friendly environment that exercise; (5) Flexibility and seamless management; (6) Attracting global players as tenants and partners; (7) A conducive operating environment. Project Overview is that Establish a Malaysian world-class transshipment port in South East Asia; Compete with the best in the world i.e. PSA of Singapore; Stop Malaysian cargo leakage to neighboring ports; Drive the development of South West Johor economy. In the aspect of Facilitation services, the port should do in this ways: (1) One-stop logistics solution for customers; (2) Simplify business set-up procedures; (3) Liaison between customers and government agencies.

Branislav Dragovic, In terms of terminal productivity, different capacity performance measures, such as berth length utilization rate (TEU/berth meter), total terminal area utilization rate (TEU/ha of total terminal area), quay crane (QC) utilization rate (TEU/QC), QC productivity (TEU/QC-hour) and average number of QCs per berth, may be used to find the optimal capacity and throughput on the terminal in port. His paper aims to make a descriptive analysis of the terminal performances interaction between ports under the framework of interaction between terminals productivity measures. Findings from this study will provide associated performance with an interesting insight into a geographical cover for the world leading port ranges. Thus, this paper develops and describes simulation

Table 2.1 Summary of Berth productivity factors

Authors	Productivity factors
Manuel Acosta	Infrastructure, superstructure, technology and communications systems
Chuan-Sheng Peng	Information technology, bigger-facilities, and automatic technology and facilities
Yang Chen	High-tech
Nam-Kyu Park	Height of container, block dispersion and distance in yard planning for loading
Kim et al.	Improve the efficient operation of port container terminal
Mu Gu	Enhanced the efficiency of loading and unloading
Carlos pestana	Technological system
Prabir De	Skilled workers, application of higher technology
Arie Sachish	Labour, capital, technology, management
Yun Cai	Berth and quay crane scheduling
Song and Cullinane	Dwell time, working time
Cullinane et al.	Service quality
Peng-fei Zhou	Berth and quay-crane assignment
Chien-Chang Chou	Yard tractor, automatic
CCDoTT	Terminal area, storage area, berth, man-hours

methodology to study the container port capacity performance. The study is based on data published on the web site of Containerization International; data were downloaded in 2005 and 2006. For each port and terminal data are referred to the waterside operation of berths and QCs are the more important determinants of productivity. We present the effect on above mentioned container terminals capacity performance with numerical results and computational experiments which are reported to evaluate the efficiency of Major European and Asian Ports (MECP and MACP).

According to CCDoTT, The productivity of container terminal can be measured with terminal's important each part, which is terminal area, storage area, berth and man-hours, etc. and selection of output has to reflects limiting component and define the fact. It emphasizes continuously monitoring of productivity with comparing with another terminal when maximum practical capacity is calculated (Table 2.1).

Chapter 3
Research Model

3.1 Establishment of Model

3.1.1 The Status of Container Ports in the World

World container throughput grew for almost 6 times in the last twelve years and reached 417 million TEUs in 2006 (Fig. 3.1). The world top six busiest ports in 2006 are the East Asian ports surrounding China, which accounts for 26 % of the world container throughput.

Over time, the use of ocean borne containers to transport international trade has affected the distribution of total maritime trade among world container ports. The increasing number of container shipments causes higher demands on the seaport container terminals, container logistics and management, as well as on technical equipment (Fig. 3.2).

Like in previous years especially Chinese ports saw a constant growth at a very high level. Even though the container traffic of the Chinese container ports grew on average by 25 % this year, the growth did not reach the last years values. All in all, the strong growth of Chinese container ports is dominating port traffic not only on a regional but on a worldwide level. Today the largest world container ports are located in Asia, especially in China and Korea.

In recent years China has poured billions of dollars into new port facilities at Shanghai, Shenzhen, Hong Kong and elsewhere. New deepwater ports and more and bigger berths have freed China from relying on foreign ports to ship its own goods.

Thanks to the development of global economy, especially the rapid expansion of Asian economy including China and Korea, and Chinese and Korean container terminal volume are steadily increasing year after year. Every port makes efforts in its own way in order to make contributions to the national economy by way of creating more value-added. At present, the China government brings forward the strategic objectives of building the international shipping center. For this study,

B. Lu and S. Wang, *Critical Factors for Berth Productivity in Container Terminal*,
DOI 10.1007/978-981-10-2431-3_3

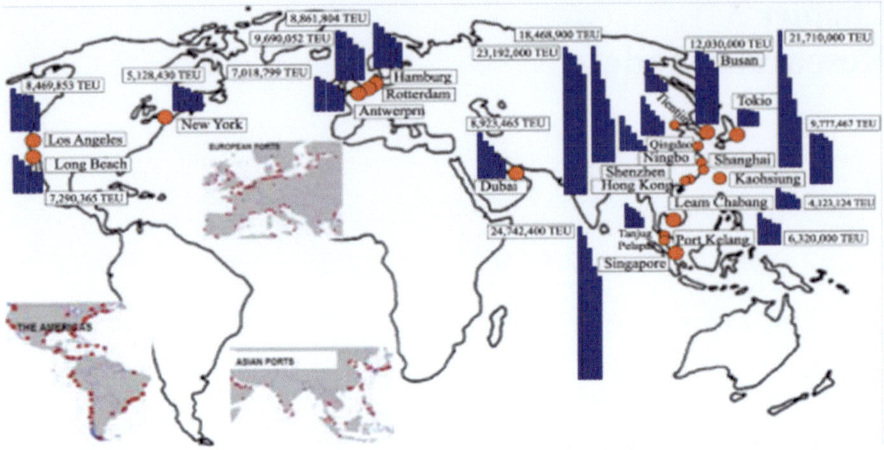

Fig. 3.1 Major world container ports in 2006

Fig. 3.2 Busan Port Boundaries

major Chinese and Korean container ports have been included in this survey in an effort to secure validity and objectivity. The target ports and container terminals surveyed are as illustrated in the Table 3.1.

The goal of this study extracts the major factors which influences to container terminal productivity and define relationship with the productivity factors and productivity.

Table 3.1 Selected Asian container terminals

Ports	Terminals Operators	Remarks
Busan	Shinsundae Jasungdae Korea Express-Gamman Hanjin Hutchison-Gamman Dongbu-Singamman Uam Terminal BICT	Direct survey
Incheon	Incheon CT	Direct survey
Gwangyang	Korea International CT Hutchison Korea terminals Kwangyang Int'l CT LDDB Crop.-Gwangyang	Direct survey
Hong Kong	HIT CSX terminal Modern terminal Cosco-HIT	Direct survey
Shanghai	Shanghai Pudong Int'l CT(SPICT) Shanghai CT(SCT)	Direct survey
Shenzhen	Shenzhen YICT Shenzhen CCT Shenzhen SCT	Direct survey

In the study applied data were gathered from 13 container terminals of Korea and 9 container terminals of China, all of which are the representative data, and also the entire container terminal which were selected are the major container terminal in the Asia and world, so the data have persuasion, potency. The selected ports will be introduced below.

3.1.2 China Ports

Port of Shanghai

The Port of Shanghai, located in the vicinity of Shanghai, comprises a deep-sea port and a river port. In 2005, with a total of 443 million tons of cargo transported, it became the world's busiest port by cargo tonnage for the first time.

The Port of Shanghai faces the East China Sea to the east, and Hangzhou Bay to the south. It includes the heads of the Yangtze River, Huangpu River (which enters the Yangtze River), and Qiantang River.

Shanghai has direct commercial contacts with over 500 ports in nearly 200 countries and regions. In 2004, Shanghai's Port ranked 3rd in the world in terms of container volume, with 14,554,000 TEUs, and 2nd in the world terms of tonnage, with 380 million tons of cargo. Enterprises from 30 countries and regions have established over 700 logistics agencies in Shanghai.

Shanghai Port is now the second largest container port in the world. Or, more accurately, it is the second in the number of standard containers handled. It reached 5.885 million standard containers in the first quarter of this year which is a rise of 28 % year on year.

Port of Hong Kong

Hong Kong, known as the fragrant harbor, has been a port for Southern China for many years. Gifted with a superb deep-water harbor offering a safe haven for ships in close proximity to the Pearl River Delta Region, Hong Kong has gradually developed over the years into a world-class container port.

(1) General Condition

Container terminals are situated in Kwai Chung-Tsing Yi basin. There are nine terminals under the operation of five different operators, namely Modern Terminals Ltd (MTL), Hong Kong International Terminals Ltd (HIT), COSCO-HIT, DP World and Asia Container Terminals Ltd (ACT). They occupy 275 hectares of land, providing 24 berths and 8,530 m deep water frontage. The water depth of the Kwai Chung-Tsing Yi basin is 15.5 m. The total handling capacity of the container terminals is over 18 million TEUs per year.

On average, some 231,000 ships, comprising both ocean vessels and river vessels for cargo and passenger traffic, visit the port of Hong Kong yearly (Table 3.2).

(2) Facilities and Equipments

The Marine Department operates and maintains 58 mooring buoys for seagoing vessels. Of these 31 are suitable for ships up to 183 m in length and 27 for ships up to 137 m. There are 44 special typhoon mooring buoys to which ships can remain secured during typhoons. This improves efficiency and reduces operational costs of vessels through elimination of unnecessary movements. In addition to the three Immigration and Quarantine Anchorages designated for visiting vessels to complete port formalities, there are eight dangerous goods and eight general-purpose anchorages providing temporary berthing spaces for vessels.

The areas and water depths of the anchorages are diversified to accommodate different sizes and draughts of ships calling at Hong Kong. There are over

Table 3.2 Facilities and equipments of Hong Kong container terminals

Terminals	Berth No.	Berth length (m)	Water depth (m)	Area (1000 × m^2)	QC No.	Management company
Total	28	9,666	16–18	177,196	153.2	
Terminal 4 (CT4)	13	4,350	16	41,446	32.7	HIT
Terminal 8 West (CT8W)	6	2,184	18	53,583	28.2	Asia Container Terminals
Terminal 5 (CT5)	7	2,432	18	40,229	38.8	Modern terminal
Terminal 8 East (CT8E)	2	700	16	41,938	53.5	Cosco-HIT

460 modern marine aids to navigation throughout Hong Kong waters to guide mariners to and from their berths.

To facilitate transfer of cargo between vessels and the shore, and cargo to and from Pearl River ports, the Marine Department provides and manages 7,756 m of public cargo-working sea frontage in various parts of Hong Kong. Bulk handling facilities for coal and oil are provided at the power generating stations at Tap Shek Kok in Castle Peak and at Po Lo Tsui on Lamma Island.

(3) Handling Performance Per Berth

In the year of 2007, Hong Kong port had increased 11 berths. Based on the year of 2008, the number of the berth is 28, the throughput is 24,494,000 TEUs, the berth length is 9,666 m, and the yard area is 177,196 square meters. The increasing tendency of throughput and the number of facilities can be found out (Table 3.3).

Distinguishing by year, the handling volumes of containers and the increase of the facilities can be distinctly found out. The handling volumes of containers of the 2006 year was more than 628,000 TEUs, the throughput which had a little decreased in 2007 year, was 768,000 TEUs. However, the throughput which was 496,000 TEUs had increased again. Therefore the high productivity can be found.

Port of Shenzhen

(1) General Condition

Shenzhen port composed by Yantian International Container Terminals (盐田), Shekou (蛇口) Container Terminals Ltd., Chiwan Container Terminals (赤湾) Ltd. The Shenzhen port, Dongguan (东莞) port and Zhuhai (珠海) port of Guangdong province developed continuously. On December, 2006, the Chinese government opened market and permitted the foreign investment for the logistics industry. Therefore, the UPS, DHL, FedEx, TNT four special express enterprises and the

Table 3.3 Throughput per berth of Hong Kong container terminals

Year	Throughput of one year (TEU)	Throughput/berth (TEU/300 m)	Throughput/berth (TEU/berth)	Throughput/berth (TEU/350 m)
2005	22,602,000	926,690	992,560	1,080,917
2006	23,230,000	952,439	1,106,190	1,110,952
2007	23,998,000	596,965	749,938	696,604
2008	24,494,000	609,303	765,438	700,705

more than 50 world-famous enterprises including Maersk had entered the Shenzhen logistics markets.

With respect to handling volumes of containers, the throughput of 2007 was 21,090,000 TEUs, up from 14.2 % 2006 year. The throughput of Shenzhen in 2008 was 21,410,000 TEUs; the growth was 1.5 % points higher over the same period last year. For throughput, Shenzhen port had been ranked 4th in the world, reduced the distance from the Hong Kong port ranked 3rd in the World, however, pulled away from the Busan port ranked 5th in the World.

The throughput of the 2003 year was 10,000,000 TEUs; however, just over a short span of four years, the throughput of the Shenzhen port had broken through 20,000,000 TEUs. The throughput of Shenzhen port had not only ten successive years ranked the 2nd in China, but also had six successive years ranked the 4th in the World. The Shenzhen port had equipped 26 container terminals. For the regular scheduled vessels arrival the port, there were 730 times in one month. The nearby Guangzhou port and the project of western development will present a threat to the existence of the throughput of Shenzhen port (Table 3.4).

(2) Facilities and Equipments

In last year, many vessels touched at Shenzhen port which increased the benefits, Shekou Container Terminals Ltd. had equipped the 2 berths in phase 2, will equip the 5 berths in phase 3. In the phase 1–3, the Yantian International Container Terminals had invested 9 berths, in phase 4, project of 1,828 m was carried out. Chiwan Container Terminals had made the plan for equipping 3 berths in phase 3. CCT will equip 4 ZPMC super post-panamax cranes.

Table 3.4 Facilities and equipments of Shenzhen container terminals

Terminals	Berth No.	Berth length (m)	Water depth (m)	Area ($1000 \times m^2$)	QC No.	Management company
Total	32	12,059	14–18	5,152	144	
Chiwan	9	3,417	14.5–16	1,246	37	CCT
Shekou	8	2,550	14–18	1,386	33	SCT
Yantian	15	6,092	16	2,520	74	YICT

Table 3.5 Throughput per berth of Shenzhen container terminals

Year	Throughput of one year (TEU)	Throughput/berth (TEU/300 m)	Throughput/berth (TEU/berth)	Throughput/berth (TEU/350 m)
2005	16,197,173	664,090	711,294	774,613
2006	18,468,900	757,232	879,471	883,257
2007	21,090,000	524,627	659,063	612,192
2008	21,410,000	532,587	669,063	612,480

(3) Handling Performance Per Berth

In the year of 2007, Shenzhen port had increased 11 berths. Based on the year of 2008, the number of the berth is 32, the throughput is 21,410,000 TEUs, the berth length is 12,059 m, and the yard area is 5,152,000 m^2. The increasing tendency of throughput and the number of facilities can be found out.

Distinguishing by year, the handling volumes of containers and the increase of the facilities can be distinctly found out. The handling volumes of containers of the 2006 year was more than 757,000 TEUs, the throughput which had a little decreased in 2007 year, was 524,627 TEUs. However, the throughput which was 532,000 TEUs had increased again. Therefore the high productivity can be found (Table 3.5).

3.1.3 Korea Port

Port of Busan

(1) General Condition

The throughput of Busan port in 2007 was 13,260,000 TEUs; the growth was 10.2 % points higher over the same period last year. The throughput of Busan port in 2008 was 13,450,000 TEUs, up from 1.4 % in 2007 year. The main reason for decreasing growth rate was the economic crisis of the American which caused the export from the Americas had decreased.

From the analytical results, after abolishing the tax of container, the inflow of the volumes of containers had increased 9.4 %. With respect to the cargo of transhipment, on April, 2007, the handling volumes of containers of MSC which had relocated from China to Busan had increased sharply which was mainly caused by the cargo of transhipment of China had increased 11.6 %.

(2) Facilities and Equipments

For expanding the facilities of Busan port, from the year of 2001, the north container terminal of construction phase 2-1 of Busan Newport had begun, and had invested 155,500,000,000 won.

From the year of 2005, the south container terminal of construction phase 2-2 of Busan Newport had begun, which will be put into use until the year of 2010. The west container terminal of construction of Busan Newport had been designed. For the super-sized vessels can come alongside the pier, especially, water depth will be dredged to 15–16 m (Table 3.6).

The berth length of Busan port was about 8.7 km, the area of Busan port was about 4,198,000 m^2. For the purpose of becoming the hub of the northeast Asia, the Busan port had done the best for investing the facilities. With the respect to throughput, different from the other Asia container ports who's the throughput had rapidly increased, especially Chinese container ports, the frequency of the handling volumes of containers of Busan port is similar with the European & American container ports.

(3) Handling Performance Per Berth

In the year of 2007, the Busan port had invested 3 berths. Base on the data of 2008 year as a standard, the number of the berth is 27, the throughput is 13,450,000 TEUs, the berth length is 8673 m, and the yard area is 4,198,000 m^2.

Distinguishing by year, the handling volumes of containers and the increase of the facilities can be distinctly found out. The handling volumes of containers of the 2005 year were more than 360,000 TEUs, the throughput which had a little

Table 3.6 Facilities and equipments of Busan container terminals

Terminals	Berth No.	Berth length (m)	Water depth (m)	Area ($1000 \times m^2$)	QC No.	Management company
Total	27	8673	11–16	4198	79	
ICT	2	700	15	298	7	BICT
Hanjin	2	600	13	142	5	HSCL
Gamman	2	700	15	305	8	HGCT, KECL
Singamman	3	826	15	308	8	DPCTCL
Sinhangman	6	2400	16	1300	18	PNPC
Sinsendae	5	1500	14–16	1012	14	PECTCL
Zasendae	5	1477	15	647	14	HCT
Uam	2	500	11	185	5	UTCL

Table 3.7 Throughput per berth of Busan container terminals

Year	Throughput of one year (TEU)	Throughput/berth (TEU/300 m)	Throughput/berth (TEU/berth)	Throughput/berth (TEU/350 m)
2005	9,124,326	366,292	380,180	427,340
2006	9,070,151	364,117	377,923	424,803
2007	9,747,184	337,156	361,007	393,349
2008	10,034,957	347,110	371,665	404,962

decreased in 2007. However, the throughput had increased again in 2008 (Table 3.7).

(a) Busan Sinsendae Container Terminal (Management Company: KBCT)

① General Condition

Last February 3rd, 22 row container cranes had been equipped. 8000 TEUs–10,000 TEUs, 5 container vessels can come alongside the pier at the same time. The total investment cost of infrastructure was about 150 billion won. In addition, the berth length had been extended to 300 m, 22 row 2 C/C and 3 RMGC has been added investments.

Based on 2009 year as the criterion, the annual handling volumes of containers of each berth are 520,000 TEUs, however the estimated optimum annual throughput are 400,000 TEUs.

② Facilities and Equipments (2008 year) (Table 3.8)

③ Throughput (2008. 1. 1–2008. 12. 31) (Table 3.9)

(b) Busan Hutchison Container Terminal (Management Company: HBCT)

① General Condition

For main line handling, 4 berths and 1 handling capacity of dock berth for 10,000 ton-class have been hold. One specialized transtainer had been equipped to handle the cargo of main line and connect the railway. From calling to vessel come alongside to the pier, the condition of more than 2 h is been recorded. Pure waiting ratio is controlled fewer than 5 %.

② Facilities and Equipments (2008 year) (Table 3.10)

③ Throughput (2008. 1. 1–2008. 12. 31) (Table 3.11)

Table 3.8 Facilities and equipments of KBCT

Berth				Total area (m^2)	Yard						
Berth length (m)	Water depth (m)	Berth No.	QC No.		Yard area (m^2)	TGS	CFS area (m^2)	Terminal crane No.			YT No.
								RTGC	RMGC	RS	
1500	14–16	5	14	1,144,588	817,592	9960	10,000	32	5	9	73

Table 3.9 Throughput of KBCT

Average vessel length (m)	Average carrying capacity (TEU)	Working time/year (day)	Working time/day (h)	Total throughput of berth (TEU)	Total throughput of yard (TEU)
236	3523	363	24	2,341,763	2,283,876

Table 3.10 Facilities and equipments of HBCT

Berth				Total area (m^2)	Yard						
Berth length (m)	Water depth (m)	Berth No.	QC No.		Yard area (m^2)	TGS	CFS Area (m^2)	Terminal crane No.			YT No.
								RTGC	RMGC	RS	
1447	10–15	5	14	647,000	462,812	10,484	19,750	32	0	2	63

Table 3.11 Throughput of HBCT

Average vessel length (m)	Average carrying capacity (TEU)	Working time/year (day)	Working time/day (h)	Total throughput of berth (TEU)	Total throughput of yard (TEU)
189	1710	363	24	2,102,969	2,101,805

(c) Busan Gamman Port (Management Company: INTERGIS)

① General Condition

The most important matter of the terminal is keeping the capacity of the marketing, in other words, keeping the mass throughput. There is big difference between singular berth and majority berths on the efficiency. The condition of the singular berth operation is more different.

② Facilities and Equipments (2008 year) (Table 3.12)

③ Throughput (2008. 1. 1–2008. 12. 31) (Table 3.13)

Table 3.12 Facilities and equipments of INTERGIS

Berth				Total area (m²)	Yard						
Berth length (m)	Water depth (m)	Berth No.	QC No.		Yard area (m²)	TGS	CFS Area (m²)	Terminal crane No.			YT No.
								RTGC	RMGC	RS	
350	15	1	4	162,750	41,777	2462	1680	10	4	5	24

Table 3.13 Throughput of INTERGIS

Average vessel length (m)	Average carrying capacity (TEU)	Working time/year (day)	Working time/day (h)	Total throughput of berth (TEU)	Total throughput of yard (TEU)
268	N/A	363	24	768,459	385,412

(d) Busan Gamchen Port (Management Company: Hanjin)

① General Condition

Gamchen container terminal is and affiliated of Hanjin, Hanjin shipping, and the business of Gamchen container terminal is carried out by planner room however not independent operation.

There are 2 handling capacity of dock berth for 50,000 ton-class. Based on 2008 year as the criterion, the annual handling volumes of containers of each berth are 280,000 TEUs which is similar with the estimated optimum annual throughput.

② Facilities and Equipments (2008 year) (Table 3.14)

③ Throughput (2008. 1. 1–2008. 12. 31) (Table 3.15)

(e) Busan Gamman Port (Management Company: HGCT)

① General Condition

Table 3.14 Facilities and equipments of Hanjin

Berth				Total area (m²)	Yard						
Berth length (m)	Water depth (m)	Berth No.	QC No.		Yard area (m²)	TGS	CFS area (m²)	Terminal crane No.			YT No.
								RTGC	RMGC	RS	
600	13.4	2	5	148,000	105,000	2238	0	12	0	1	23

Table 3.15 Throughput of Hanjin

Average vessel length (m)	Average carrying capacity (TEU)	Working time/year (day)	Working time/day (h)	Total throughput of berth (TEU)	Total throughput of yard (TEU)
233	2661	363	24	559,138	544,725

Formerly, there is the bottleneck phenomenon that the occupancy rate of the container yard was only 43 %. However, after improving the operation system, the occupancy rate of the container yard can reach 70 %.

The productivity of the container port is depended on productivity of crane and Turn Around Time. Moreover, for the handling capability, the affection of container yard is absolute small.

Transshipment ratio is about 40 %, which is slow increasing. The cargo is almost from local cargo. The condition of degeneration is caused by the Shanghai yangshan container terminal.

② Facilities and Equipments (2008 year) (Table 3.16)

③ Throughput (2008. 1. 1–2008. 12. 31) (Table 3.17)

(f) Busan Singamman Port (Management Company: DPCT)

① General Condition

The shape of the container yard is not triangle, however is diamond-shaped. The application of nearby region is difficult, because of the implied land cost, the efficiency had dropped sharply.

For the condition of the single kind of berth, the optimum occupancy rate of the container yard is about 70 %. The optimum once handling capacity is controlled by 500 TEUs–1000 TEUs.

② Facilities and Equipments (2008 year) (Table 3.18)

③ Throughput (2008. 1. 1–2008. 12. 31) (Table 3.19)

(g) Busan Uam Port (Management Company: UTC)

① General Condition

Table 3.16 Facilities and equipments of HGCT

Berth				Total area (m^2)	Yard						
Berth length (m)	Water depth (m)	Berth No.	QC No.		Yard area (m^2)	TGS	CFS area (m^2)	Terminal crane No.			YT No.
								RTGC	RMGC	RS	
350	15	1	4	149,000	129,408	2994	1705	10	0	3	23

Table 3.17 Throughput of HGCT

Average vessel length (m)	Average carrying capacity (TEU)	Working time/year (day)	Working time/day (h)	Total throughput of berth (TEU)	Total throughput of yard (TEU)
243	3492	363	24	650,570	650,994

Table 3.18 Facilities and equipments of DPCT

Berth				Total area (m²)	Yard						
Berth length (m)	Water depth (m)	Berth No.	QC No.		Yard area (m²)	TGS	CFS area (m²)	Terminal crane No.			YT No.
								RTGC	RMGC	RS	
826	15	2.5	7	308,000	153,490	4374	5500	17	7	3	36

Table 3.19 Throughput of DPCT

Average vessel length (m)	Average carrying capacity (TEU)	Working time/year (day)	Working time/day (h)	Total throughput of berth (TEU)	Total throughput of yard (TEU)
195	2135	363	24	1,227,494	1,143,718

UTC is established by KCTC and Goo bao, therefore the two companies' data can be combined and arrangement.

The reasonable occupancy rate of terminal yard is 60 %. If the occupancy rate exceeds 60 %, the normal operation is different.

② Facilities and Equipments (2008 year) (Table 3.20)

③ Throughput (2008. 1. 1–2008. 12. 31) (Table 3.21)

(h) Busan Gamman Port (Management Company: BICT)

① General Condition

In October, 2005, the rented room in Gamman port cooperated with Hanjin shipping. The information system had organized again. The previous data had also combined.

Table 3.20 Facilities and equipments of UTC

Berth				Total area (m²)	Yard						
Berth length (m)	Water depth (m)	Berth No.	QC No.		Yard area (m²)	TGS	CFS area (m²)	Terminal crane No.			YT No.
								RTGC	RMGC	RS	
500	11	1.5	5	185,000	156,000	2627	0	13	0	2	24

Table 3.21 Throughput of UTC

Average vessel length (m)	Average carrying capacity (TEU)	Working time/year (day)	Working time/day (h)	Total throughput of berth (TEU)	Total throughput of yard (TEU)
141	938	363	24	569,735	592,185

Table 3.22 Facilities and equipments of BICT

Berth				Total area (m²)	Yard						
Berth length (m)	Water depth (m)	Berth No.	QC No.		Yard area (m²)	TGS	CFS area (m²)	Terminal crane No.			YT No.
								RTGC	RMGC	RS	
700	15	2	7	297,535	280,000	6.571	1300	21	7	4	34

② Facilities and Equipments (2008 year) (Table 3.22)

③ Throughput (2008. 1. 1–2008. 12. 31) (Table 3.23)

Port of Gwangyang

① General Condition

The average occupancy rate of the berth is 40.2 %. Because of accepting the demand of long-term free equipment from the large shipping company, the Yard Density is average exceed 83 %. Therefore, the space of the container yard is insufficient. The temporary empty containers are loaded outside the grounds.

The handling capacity of 50,000 ton-class 1 berth is operating. Because the container yard is small, a fair number of empty containers are being transported to the suburb; however, it need afford costs upon them.

According to the analysis results, the yard density is exceeding 60 %. The prediction of cargo handling is difficult which is caused by the unreasonable demand that the free loading time exceeds 30 days.

The World No. 1 ranking Maersk shipping company, comparing with the others, is handling the large cargo in Gwangyang port.

② Facilities and Equipments (2008 year) (Table 3.24)

③ Throughput (2008. 1. 1–2008. 12. 31) (Table 3.25)

(c) Gwangyang HKTL Phase 1

① General Condition

The unreasonable demand that the free loading time exceeds 30 days from the large shipping company caused the yard density is high. The service of on dock is operating. When the contract is expired, the problems often arise.

Table 3.23 Throughput of BICT

Average vessel length (m)	Average carrying capacity (TEU)	Working time/year (day)	Working time/day (h)	Total throughput of berth (TEU)	Total throughput of yard (TEU)
240	3625	363	24	1,265,993	1,254,316

Table 3.24 Facilities and equipments of KOREX

Berth				Total area (m^2)	Yard						
Berth length (m)	Water depth (m)	Berth No.	QC No.		Yard area (m^2)	TGS	CFS area (m^2)	Terminal crane No.			YT No.
								RTGC	RMGC	RS	
350	15	1	3	210,000	140,700	3726	4080	6	0	2	12

Table 3.25 Throughput of KOREX

Average vessel length (m)	Average carrying capacity (TEU)	Working time/year (day)	Working time/day (h)	Total throughput of berth (TEU)	Total throughput of yard (TEU)
N/A	N/A	363	20	403,603	142,906

The yard density is more than 60 %, in the ordinary day; the condition of low productivity is occurrence.

② Facilities and Equipments (2008 year) (Table 3.26)

③ Throughput (2008. 1. 1–2008. 12. 31) (Table 3.27)

(d) Gwangyang GICT Phase 1

① General Condition

The Yard Density is exceeding 40 %. Moreover, competitive cargo attraction and the unreasonable demand that the free loading time exceeds 30 days from the large shipping company cause the optimum maintaining is difficulty.

Table 3.26 Facilities and equipments of HKTL

Berth				Total area (m^2)	Yard						
Berth length (m)	Water depth (m)	Berth No.	QC No.		Yard area (m^2)	TGS	CFS area (m^2)	Terminal crane No.			YT No.
								RTGC	RMGC	RS	
350	16	1	2	210,000	125,000	3630	4800	3	0	3	15

Table 3.27 Throughput of HKTL

Average vessel length (m)	Average carrying capacity (TEU)	Working time/year (day)	Working time/day (h)	Total throughput of berth (TEU)	Total throughput of yard (TEU)
191	2743	363	24	124,590	125,111

Table 3.28 Facilities and equipments of GICT

Berth				Total area (m^2)	Yard						
Berth length (m)	Water depth (m)	Berth No.	QC No.		Yard area (m^2)	TGS	CFS area (m^2)	Terminal crane No.			YT No.
								RTGC	RMGC	RS	
700	15	2	5	420,000	396,600	26,061	8640	12	0	5	23

Table 3.29 Throughput of GICT

Average vessel length (m)	Average carrying capacity (TEU)	Working time/year (day)	Working time/day (h)	Total throughput of berth (TEU)	Total throughput of yard (TEU)
N/A	N/A	363	20	152,239	150,689

② Facilities and Equipments (2008 year) (Table 3.28)

③ Throughput (2008. 1. 1–2008. 12. 31) (Table 3.29)

(e) Gwangyang GDBP Phase 2-1

① General Condition

The optimum Yard Density is 45 %, in the ordinary day, the condition of low productivity is occurrence.

② Facilities and Equipments (2008 year) (Table 3.30)

③ Throughput (2008. 1. 1–2008. 12. 31) (Table 3.31)

Table 3.30 Facilities and equipments of GDBP

Berth				Total area (m^2)	Yard						
Berth length (m)	Water depth (m)	Berth No.	QC No.		Yard area (m^2)	TGS	CFS area (m^2)	Terminal crane No.			YT No.
								RTGC	RMGC	RS	
350	16	1	2	206,984	125,000	3712	3636	4	0	1	15

Table 3.31 Throughput of GDBP

Average vessel length (m)	Average carrying capacity (TEU)	Working time/year (day)	Working time/day (h)	Total throughput of berth (TEU)	Total throughput of yard (TEU)
150	994	363	24	166,371	165,802

Port of Incheon

(a) Incheon ICT

① General Condition

Because the sea route is narrow and the yard area is small, the yard density is exceeding 75 %, and the total handling volumes will break 400,000 TEUs. The demand that the free loading time exceeds 30 days from the large shipping company is being operated.

② Facilities and Equipments (2008 year) (Table 3.32)

③ Throughput (2008. 1. 1–2008. 12. 31) (Table 3.33)

(b) Incheon Sun Kwang

① General Condition

The behind open-air storage yard is being applied by the container yard.

② Facilities and Equipments (2008 year) (Table 3.34)

③ Throughput (2008. 1. 1–2008. 12. 31) (Table 3.35)

Table 3.32 Facilities and equipments of ICT

Berth				Total area (m^2)	Yard						
Berth length (m)	Water depth (m)	Berth No.	QC No.		Yard area (m^2)	TGS	CFS area (m^2)	Terminal crane No.			YT No.
								RTGC	RMGC	RS	
600	14	2	6	137,771	113,067	2482	0	8	0	3	14

Table 3.33 Throughput of ICT

Average vessel length (m)	Average carrying capacity (TEU)	Working time/year (day)	Working time/day (h)	Total throughput of berth (TEU)	Total throughput of yard (TEU)
N/A	1073	363	24	344,896	334,896

Table 3.34 Facilities and equipments of SunKwang

Berth				Total area (m^2)	Yard						
Berth length (m)	Water depth (m)	Berth No.	QC No.		Yard area (m^2)	TGS	CFS area (m^2)	Terminal crane No.			YT No.
								RTGC	RMGC	RS	
407	11	2	3	244,546	25,327	3690	2170	2	4	1	8

Table 3.35 Throughput of SunKwang

Average vessel length (m)	Average carrying capacity (TEU)	Working time/year (day)	Working time/day (h)	Total throughput of berth (TEU)	Total throughput of yard (TEU)
164	776	363	20	29,543	29,545

3.2 Definition of Terms in Container Terminal

3.2.1 *Throughput*

In container terminal, throughput is the amount of work that a terminal can do in a given time period. Historically, throughput has been a measure of the comparative effectiveness of large terminals that deal with many containers concurrently. An early throughput measure was the number of TEUs deal withed in one day, one month, and one year.

Another measure of container terminal throughput is performance, the speed with which time one or a set of containers facilities run with a certain workload or how many containers are being handled in a fixed time. A benchmark can be used to measure throughput.

3.2.2 *Container Terminal Facilities*

Berth

There are many definitions about berth as follows; First sufficient space for a ship to maneuver; sea room: kept a clear berth of the reefs Second, s space for a ship to dock or anchor: (a) steamship moored to its berth at the pier. Third, employment on a ship: sought an officer's berth in the merchant marine. fourth, sabuilt-in bed or bunk, as on a ship or a train. (b) A place to sleep or stay; accommodations: found a berth in a nearby hotel. Fifth, a space where a vehicle can be parked, as for loading. In another words, container terminal berth is the requirement length of shore when the vessels berth. The length of berth is different according to the size of vessels. In a general way, the length of berth is 250 and 350 m.

Container Yard

Container Yard, CY in short, in a general way has two meanings: from the angle of broad sense, container yard can be understood that the place which assembles and unassembled the containers, hand over and take over containers, storing the containers, including Marshalling Yard, Container Yard and shore.

Marshalling Yard is the place which before the container vessels importing, container operators should stack the containers well which will export according to instruction sheet; when unloading the import containers, arrange the containers on the container yard with the request and sequence of consignment, for the purpose of arranging to obligate the place just is Marshalling Yard.

Container Yard is the yard which is no direct relation with the stack and vessels assembling and unassembling operation. In a general way, container yard is far from the container yard, even, the container yard is out side extension of port area.

3.2.3 Container Terminal Equipments

Quay Crane

Quay crane is the main equipment to play the core role of loading and unloading containers in seaport terminal. Besides it possesses the interface function between waterside and stevedoring systems, the performance of quay cranes is naturally one of the most important indexes for assessing the efficiency of a container terminal. The problem of quay crane scheduling not only considers the arrangement of service time windows, but also needs to avoid crossing each other and to assign neighbored cranes for a vessel.

In Container Yard Equipments

Both overhead traveling cranes and gantry cranes are types of crane which lift objects by a hoist which is fitted in a trolley and can move horizontally on a rail or pair of rails fitted under a beam. An overhead traveling crane, also known as an overhead crane or as a suspended crane, has the ends of the supporting beam resting on wheels running on rails at high level, usually on the parallel side walls of a factory or similar large industrial building, so that the whole crane can move the length of the building while the hoist can be moved to and fro across the width of the building. A gantry crane has a similar mechanism supported by uprights, usually with wheels at the foot of the uprights allowing the whole crane to traverse.

Overhead traveling cranes and gantry cranes are particularly suited to lifting very heavy objects and huge gantry cranes have been used for shipbuilding where the crane straddles the ship allowing massive objects like ships' engines to be lifted and moved over the ship. Two famous gantry cranes built in 1974 and 1969 respectively are Samson and Goliath, which reside in the largest dry dock in the world in Belfast, Northern Ireland. Each crane has a span of 140 m and can lift loads of up to 840 tones to a height of 70 m, making a combined lifting capacity of over 1600 tones, one of the largest in the world.

However, gantry cranes are also available running on rubber types so that tracks are not needed, and small gantry cranes can be used in workshops, for example for lifting automobile engines out of vehicles.

RTGC (Rubber Tyred Gantry Crane)

A Rubber Tyred Gantry Crane is typically used for movement of containers and loading of trucks. This crane type usually consists of three separate motions for transportation of material. The first motion is the hoist, which raises and lowers the material. The second is the trolley gear, which allows the hoist to be positioned directly above the material for placement. The third is the gantry, which allows the entire crane to be moved along the working area.

Special features available:

- Hoisting height up to 1-over-6 containers.
- 8 containers wide plus truck lane.
- 50.8 tones/50 long tons twin lift.
- 65 tones/64 long tons twin lift.
- Liebherr DC drive control systems.
- DGPS automatic gantry steering.
- DGPS automatic container position recognition.

RMGC

Rail-mounted gantry crane (RMG) is a specialized yard container handling machine. It can travel on rail by means of the yard power, and lift and stack containers in the yard area with the equipment of the 20' or 40' telescopic spreader (or twin-lift spreader if needed). Compared with the rubber-tyred gantry crane (RTG), the RMG has the advantages of being driven by electrical power, clean, bigger lifting capacity, and high gantry traveling speed with cargo.

The RMG consists of lifting mechanism, trolley traversing mechanism, gantry mechanism and sway-dampening mechanism.

The lifting, gantry and trolley mechanisms are mostly equipped with AC frequency conversion control system. Normally the lifting mechanism is of the single drum type. It can also be designed to be double drum type if needed.

The following items are from users' selection according to their own needs:

- Trolley slewing mechanism. Normally the trolley can be slewed easily between $-10°$ and $+190°$. If needed, continuous slewing can also be realized.
- RMGS with overhanging en beam, which reaches beyond gantry traveling rail.
- RMGS with different rail gauges.
- RMGS with generator.
- Two-direction anti-sway system (eight-rope system, refer to left).

RMCS

The design of the rail-mounted container stacking cranes responds to the individual customer requirements. Features include:

- High Hoist and Trolley speeds resulting in short cycle time.
- Simultaneous Hoisting, Trolley Traveling and Gantry Travel as standard.
- Twin lift Operation up to 65 Tones. (2 × 32.5 Tones).
- The latest Speed Control Systems (Liebherr AC and DC Drive Systems).
- Sophisticated Crane Management System providing extensive operational and crane performance data.
- Proven No-Sway System.
- High reliability/availability figures with MTBF values of less than 0.5 % per annum.
- Fault Monitoring/Maintenance Support direct from the Liebherr factory via RF Link/LAN and Modem.
- Stable, rigid, sway resistant, high quality steel structure design.

3.3 Setting up Concept of Container Terminal Productivity

Throughout research review about the concept of productivity, e.g. Risaburo Nezu, and the method of productivity measurement. productivity is commonly defined as a ratio of a volume of output to a volume of inputted resources. While there is no disagreement on this general notion, a look at the productivity literature and its various applications reveals that there is neither a unique purpose for, nor a single measure of, productivity. Productivity in economics refers to measures of output from production processes, per unit of input. Labor productivity, for example, is typically measured as a ratio of output per labor-hour, an input. Productivity may be conceived of as a measure of the technical or engineering efficiency of production. As such quantitative measures of input, and sometimes output, are emphasized. Productivity is distinct from measures of allocation efficiency, which take into account both the value of what is produced and the cost of inputs used, and also distinct from measures of profitability, which address the difference between the revenues obtained from output and the expense associated with consumption of inputs. Thus, the productivity is applied with the important judgment of the efficiency measurement.

There are many different productivity measures. The choice between them depends on the purpose of productivity measurement and, in many instances, on the availability of data. Broadly, productivity measures can be classified as single factor productivity measures (relating a measure of output to a single measure of input) or multi factors productivity measures (relating a measure of output to a bundle of inputs). Another distinction, of particular relevance at the industry or firm level is between productivity measures that relate some measure of gross output to one or several inputs and those which use a value-added concept to capture movements of output.

Accordingly, an increase in productivity is characterized by a shift of the production function and a consequent change to the output/input relation. The formula of total productivity is normally written as follows in Formulation-1:

$$\text{Total productivity} = \text{Output quantity}/\text{Input quantity} \qquad \text{(Formulation-1)}$$

So far, most of researches which relate to port productivity have focused on the ratio of annual handling capacity and the spent resources in the container terminal. This means productivity is depend on some aspects like marketing power, agility of business organization, the terminal area and the scale of national economy. In this point, there are a lot of rooms for considering in handling productivity problem in comparison with in handling unit facility productivity.

In the viewpoint of the efficiency which get a larger output with a smaller input, the productivity of the container terminal is defined as a volume of loading and unloading per unit time per spent resources and can be expressed with relationships which sets the volume of loading and unloading as dependent variable and the inputted resources as independent variables.

This study handles gross berth productivity (GBP) as productivity indicator. The function of the harbor converted from military affairs, fishery and the passenger service center to import and export freight center in the international trade. So, the ability for loading and unloading is most important in a port operation and it can be measured with the volume of loading and unloading per unit resource. The volume of loading and unloading per unit resource is set with annual throughput per berth i.e. with annual throughput divided by total berthing time and number of berths.

3.4 Scope of Modeling

In order to determine productivity factors in a container terminal, all factors that are relevant to terminal operation, are to be considered such as terminal facilities like yard area, number of berth, gate, and rail station, and terminal equipment like YT, QC, RTGC, RMGC, reach stacker, top handler and folk lifter and manpower.

In addition to these facilities and equipment, worker skill, the level of information technology, the extent of terminal automation and operating time is closely

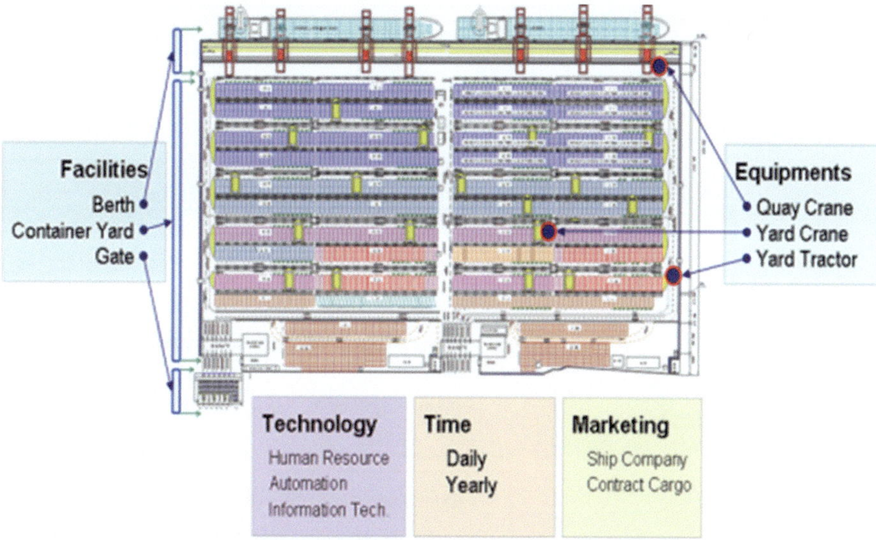

Fig. 3.3 Productivity factors in container terminal

related to productivity. Figure 3.3 simply shows related factors which affecting container terminal productivity. This figure is drawn from South Korea's main terminal in 2006, based on operational data.

3.5 Modeling Method

Independent variables are established by extracting the whole major factors which have relationship with productivity and by normalization process. For the purpose of considering the potential factors, it is necessary to do literature review, web site survey, interview brain-storming, and checking duplicated factors in context.

Normalization process consists of three steps. First the extracted elements which were in a disorder are categorized by general criteria on facility, equipment, technology, business activity and working time. Next step is to find duplicated factors in comparing the attribute context each other. If duplication in context is found, the factor is excluded from independent factors. Final step is to apply other criteria to remaining factors in order to identify if it has qualification of independent variables.

After final selection of the independent factors, interrelationship between independent variables and dependent variables can be established (Fig. 3.4). After collecting data, regression analysis is done by procedure such as enter, remove, stepwise, forward and backward, with using GBP as dependent variable.

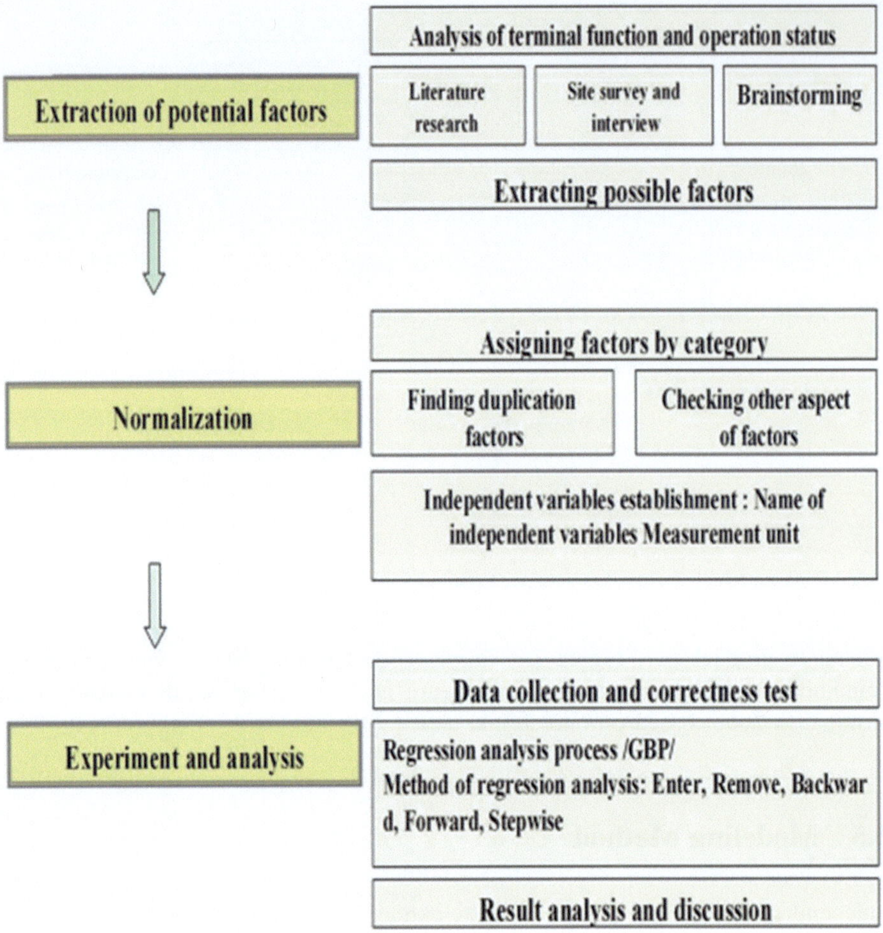

Fig. 3.4 Research procedure

3.6 Method of Regression Analysis

In this study, we use the SPSS software to analyze the data and make regression analysis. In research, the regression analysis method is used for the modeling and analysis of numerical data consisting of values of a dependent variable (response variable) and of one or more independent variables (explanatory variables). The dependent variable in the regression equation is modeled as a function of the independent variables, corresponding parameters ("constants"), and an error term. The error term is treated as a random variable. It represents unexplained variation in the dependent variable. The parameters are estimated so as to give a "best fit" of the data. Most commonly the best fit is evaluated by using the least squares method, but other criteria have also been used.

The goal of regression analysis is to determine the values of parameters for a function that cause the function to best fit a set of data observations that you provide. In linear regression, the function is a linear (straight-line) equation.

Linear regression is a form of regression analysis in which observational data are modeled by a function which is a linear combination of the model parameters and depends on one or more independent variables. In simple linear regression the model function represents a straight line. The results of data fitting are subject to statistical analysis.

The various variable selection methods are applied on the ground of 'principle of parsimony' which removes independent variable of coherence low-end in a multiple regression model.

In statistics, stepwise regression includes regression models in which the choice of predictive variables is carried out by an automatic procedure. Usually, this takes the form of a sequence of F-tests, but other techniques are possible, such as t-tests, Adjusted R-square, Akaike information criterion, Bayesian information criterion, Mallows' Cp, or false discovery rate.

The main approaches are as follows:

(a) Forward selection, which involves starting with no variables in the model, trying out the variables one by one and including them if they are 'statistically significant'.
(b) Backward selection, which involves starting with all candidate variables and testing them one by one for statistical significance, deleting any that are not significant.
(c) Methods that are a combination of the above, testing at each stage for variables to be included or excluded.

A widely used algorithm was proposed by Efroymson. This is an automatic procedure for statistical model selection in cases where there are a large number of potential explanatory variables, and no underlying theory on which to base the model selection. The procedure is used primarily in regression analysis, though the basic approach is applicable in many forms of model selection. This is a variation on forward selection. At each stage in the process, after a new variable is added, a test is made to check if some variables can be deleted without appreciably increasing the residual sum of squares (RSS). The procedure terminates when the measure is (locally) maximized, or when the available improvement falls below some critical value.

This study try to use different methods such as Enter, Remove, Backward, Forward, Stepwise, and then Backward and to select the best method in terms of results.

After test of the regression, we found that backward elimination give better result container terminals in Busan City and forward selection give better result container terminals in Gwangyang City.

Backward elimination method progress as follows. Firstly, estimate the model with all the predictors. If all the regression coefficients are significant, the procedure stops. Secondly, the predictor with the smallest significance in the regression coefficient test is eliminated from the model. Finally, the procedure stops, when all the regression coefficients are significant. On the one hand, forward selection method progress as follows. Firstly, the variable, which has the strongest and significant correlation, is taken into the model. Secondly, calculate the partial correlation between the variables not in the equation and the model. The variable with the strongest and significant correlation is taken into the model. Finally, the procedure continues until there is no significant partial correlation. The statistical analysis of GBP is implemented in two types because the relative difference of data between port of Busan and port of Gwangyang is remarkable. For Chinese ports, because of no big difference, so the statistical analysis of Chinese ports GBP in only one type.

3.7 Deciding Independent Variables

Referring to deciding independent variables, I want to cite the N.K. Park et al. research result. There are various general factors impacting terminal productivity, which can be distinguished from facility, equipment, technology, business activity and working time.

In considering facility related to factors, we can extract berth factor where vessels lie alongside, the container yard factor where containers stock, the gate factor where containers enter and leave. In considering the berth related factor in detail, we can extract the number of berth, the length of berth, and berth occupancy ratio as independent factor. However as the number of berth have duplicate characteristics with the length of berth, it can be excluded as independent factors. Furthermore berth occupancy ratio can be excluded, because it is not the input factor of productivity but a result of terminal operation.

In considering container yard related factor, the total terminal area, TGS, the height of stack, dwell time, and utilization of container yard can be extracted as major factors of productivity, yet TGS is excluded because it can be duplicated with the total area. Furthermore, the height of container stack in container yard, even if they are the most important factors, is also excluded because its significance is too low to be guarantee. Container yard utilization ratio is excluded because it is the result of container terminal operation rather than input factor.

Other factors such as the number of lanes at gate, automation technology at gate or on yard are to be considered. Among those factors, the difficulty to measure the level of automation technology prevents it from getting qualification as independent factors.

Let me say equipment factors as productivity indicator. Quay crane, yard crane, and transfer equipment are most important productivity factors. Referring to equipment, two factors are to be listed. The one is how many of QC, TC, YT do each terminal own for handling operating. The other is about the lifting capability of equipment. Particularly, there are many types in QC in respect of arm's length reaching to 24 bays on container vessel, QC with twin lift which has capability to lift two 20 ft container all together, QC's height etc. Furthermore, on yard operation, RMGC shows higher productivity in handling rather than RTGC, and reach stacker, top handler, straddle carrier, folk lifter which container terminal owns shows different performance by terminal. In yard tractor operation, different performances are revealed depending on operation type like YT pooling or dual cycle. However, in this paper, I assume the same type of equipment has similar performance on average. As a result of discussion, the number of equipment is only factor to consider for productivity factor.

In referring to operation technology, big variance in performance exists depending on workers skill, the level of information technology and automation in terminal. If we analyze workers' skill in detail level, it can be divided into skill of site work like QC, YT, TC drivers, and management skill on level of operation, tactical and strategy. However, measuring workers and management skill is not only difficult to ensure objective validity, but also estimate the impact to productivity on worker efficiency.

Referring to the level of information technology, different types of technology are implemented on gate, on yard operation and on quay side. On gate, yard and quay in order to get identification of truck or container, barcode system, optical recognition system, RFID recognition system are selected according to container terminal's budget and manager preference to recognition accuracy. Referring to terminal operation system, EDI system which links to shipping company and terminal operator, terminal operation system called TOS which has control over terminal operation are implemented by all terminals. Referring to automation, unmanned yard crane used in PECT is only example in Korea even though automation in horizontal and yard handling will be implemented on southern part of Busan New Port. However, measuring the level of information and automation technology is not only difficult to ensure objective validity, but also estimate the impact to productivity on technology level.

Finally it is time to discuss business activity of Terminal Operation Company (TOC) regarding to the number of contracted shipping company and contracted amount of cargo. On high level view, even if we accept that TOC's business activity has influenced to annual GBP, measuring the level of business activity regarding to productivity is not only difficult to collect data owing to confidential property. Nevertheless, it is true that working hours affects productivity, but it can not be applied in this study because it has no difference by each terminal.

The above discussion will be summarized on Table 3.36 reflecting context duplication characteristic and other aspect of factors.

Table 3.36 Independent variables

Class	Resource name	Factor	Unit	Source	Review
Facilities	Berth	Berth Number	ea	(4)	Duplication
		Berth Length	**m**	**(4), (5)**	**Adopted**
		Berth Occupancy	%	(1), (2), (3)	Result of operation
	CY	**CY Area**	**m²**	**(4), (5)**	**Adopted**
		Stack Capacity	TGS	(4), (5)	Duplicated
		Stack Height	tier	All	Insufficient significance
		CY Dwell Time	**day**	**All**	**Adopted**
		Land Utilization	%	(1), (3)	Result of operation
	Gate/Rail	Number of lanes	ea	(4), (5)	Assume similar
Equipments	Q/C	**Number**	ea	**All**	**Adopted**
		Capability	TEU/h	(1), (2), (3)	Assume similar performance
	Y/C	**Number**	ea	**(4), (5)**	**Adopted**
		Capability	TEU/h	(4), (5)	Assume similar performance
	Y/T	**Number**	**Ea**	(4), (5)	**Adopted**
		Capability	TEU/h	(4), (5)	Assume similar performance
Technology	IT	High	N/A	(4), (5)	Difficult measurement
		Medium	N/A	(4), (5)	Difficult measurement
		Low	N/A	(4), (5)	Difficult measurement
	Automatic	High	N/A	(4), (5)	Difficult measurement
		Medium	N/A	(4), (5)	Difficult measurement
		Low	N/A	(4), (5)	Difficult measurement
	Human Resource	High	N/A	(4), (5)	Difficult measurement
		Medium	N/A	(4), (5)	Difficult measurement
		Low	N/A	(4), (5)	Difficult measurement
Business activity	Contract Amount of Cargo		TEU/year	(4), (5)	Difficult measurement
	Contract Shipping Company		ea/year	(4), (5)	
Working time	Working Time per day		h	All	No difference
	Working Time per year		days	All	

Source (1) UNCTAD, (2) Hamburg Port Consulting, (3) Korea Maritime & Fisheries Institute, (4) Interview, (5) Papers which previously referred

Table 3.37 Korea Data for GBP

Class	Dependent Var. and related factors						Independent Var.										
	GBP (dependent)	Related factors					Facility							Equipment			
		Thr.	Total Berthing time	Berth Occ. (%)	Working time		Berth		Yard				Gate	Q/C	T/C		Y/T
					Day	Year	No	Length	TGS	Area	Height of stack	Dwell time	Lane	Q'ty	R T GC	RM GC	Q'ty
															Q'ty	Q't	
PUS-1	109.83	1,847,950	3739	52	20	363	4.5	322	2330	143,778	3.5	5.7	1.3	3.1	7.6	0.2	13.3
PUS-2	115.29	2,008,772	4356	60	20	363	4.0	300	2490	204,398	2.6	5.3	2	3	10	1	18
PUS-3	168.65	662,398	3928	54	20	363	1.0	350	2994	129,408	4.0	3.0	4	4	13	0	23
PUS-4	168.10	976,319	5808	80	20	363	1.0	350	2462	41,777	4.0	3.0	3	4	15	4	24
PUS-5	138.77	1,350,000	4864	67	20	363	2.0	350	3286	148,677	3.5	3.0	3	4	13	4	17
PUS-6	88.87	1,000,000	4501	62	20	363	2.5	330	1750	123,200	3.5	7.0	3	3	8	3	14
PUS-7	99.33	533,623	2686	37	20	363	2.0	300	1119	42,735	4.0	3.2	2	3	7	0	12
PUS_8	58.84	598,000	5082	70	20	363	2.0	250	1314	92,500	3.5	3.0	3	3	8	0	12
KW-1	107.26	124,590	1162	16	20	363	1.0	350	3630	210,000	3.0	0.0	6	2	6	0	15
KW-2	28.23	258,192	1307	18	20	363	7.0	279	1969	61,830	4.5	2.0	1	1	1	1	3
KW-3	67.40	166,371	2468	34	20	363	1.0	350	3712	125,000	4.0	5.0	6	2	5	0	15
KW-4	21.84	152,239	3485	48	20	363	2.0	350	13,031	198,300	3.0	7.0	3	3	9	0	12
Other-1	89.63	344,896	3848	53	20	363	1.0	300	2482	113,067	4.5	15.0	4	3	11	0	14
Other-2	20.35	29,543	726	10	20	363	2.0	204	639	11,454	4.0	27.0	2	2	2	2	4
Other-3	40.19	40,850	1016	14	20	363	1.0	240	1496	96,000	4.0	4.1	2	2	6	0	10
Other-4	8.97	18,229	1016	14	20	363	2.0	240	186	860			1	2	1	0	0

Table 3.38 Major productivity factors in Busan and Kwangyang

Class	Busan	Gwangyang
GBP	58.8–168.6	21.8–107.2
Berth occupancy	37–80 %	16–48 %
Length of berth	250–350	279–350
Yard area	41,777–204,398	61,830–210,000
Gate lane	1.3–4	1–6
Q/C	3–4	1–3
T/C	7–19	2–9
Y/T	12–24	3–15
Dwell time	3–15	0–7

3.8 Description of Korea Data

The data applied in this study are obtained from the year of 2006. For Korea ports, Eight terminals located in Busan, four containers in Gwangyang, one terminal in Incheon and two terminals in other area in Korea. As two terminals in Incheon and other city among target terminals just opened, they are excluded from analytical process because it does not keep normal operation. Finally, thirteen terminals data can be actually applied for analysis, which are eight in Busan, four in Gwangyang, and one in Incheon included in Busan data.

For making model, GBP per berth can be set as dependent variable. As we already discussed, number of berths, height of stack in the independent variables are excluded owing to lower significance after regression analysis as shown Table 3.36. Finally the length of berth, yard area, dwell time, number of quay crane, yard crane and yard tractor are chosen with strong reliability as independent variables.

Through collecting the Korea data and we can conclude and get the Korea data for GBP Table 3.37.

In summary for collected data, distribution analysis is useful to catch whole figure of target terminals resource. Table 3.38 shows resource distribution of terminal both in port of Busan and port of Gwangyang by gross berth productivity.

As shown above Table 3.38, the remarkable GBP difference between in port of Busan and port of Gwangyang is found through the result of data analysis. If data both in port of Busan and port of Gwangyang put into one basket together, the 'error of combinations' can be happened due to wide variance. Therefore the statistical analysis is implemented separately by port of Busan and port of Gwangyang.

3.9 Description of China Data

The data applied in this study are obtained from the year of 2006. Two terminals located in Shanghai, four containers in Hong Kong, three terminals in Shenzhen can be actually applied for analysis.

Table 3.39 China Data for GBP

Class	Dependent Var. and related factors						Independent Var.									
	GBP (dependent)	Related factors					Facility							Equipment		
		Thr.	Total Berthing time	Berth Occ. (%)	Work Time		Berth		Yard				Gate	Q/C	T/C	Y/T
					Day	Year	No.	Length	TGS	Area	Height stack	Dwell time (days)	Lane	Q'ty	Q'ty	Q'ty
SHA-1	225.27	14,800,000	6570	75	24	365	10	3450	3450	54,589	4	1	2	2.0	5.4	11.7
SHA-2	234.74	2,900,000	4118	47	24	365	3	900	7233	71,900	3	3	1	4.0	14.0	24.3
SHE-1	140.46	8,860,000	6308	72	24	365	10	3380	2629	50,000	3.5	1	3	4.4	7.8	24.6
SHE-2	113.72	4,224,000	4643	53	24	365	8	2,917	3375	19,750	3	1	2	5.1	6.4	12.6
SHE-3	139.74	5,730,000	4556	52	24	365	9	3050	3037	13,333	3	1	1	3.0	10.3	12.6
HK-1	135.01	12,300,000	7008	80	24	365	13	4350	2718	41,446	4.5	0.5	3	4.6	11.6	16.5
HK-2	251.8	9,000,000	5957	68	24	365	6	2,184	3576	53,583	3.5	0.5	2	5.7	4.7	17.8
HK_3	172.26	7,500,000	6220	71	24	365	7	2,432	4048	40,229	3	0.5	2	4.3	15.1	19.4
HK-4	227.22	2,150,000	4731	54	24	365	2	700	4000	41,938	3	5.5	3	5	16	32.5

Table 3.40 Major productivity factors in Chinese port

Class	China
GBP	113.72–251.8
Berth Occupancy	47–80 %
Length of berth	335–350
Yard area	13,333–71,900
Gate lane	1–3
Q/C	1.3–5.7
T/C	4–16
Y/T	8.1–32.5
Dwell time	0.5–3

Also, through collecting the China data and we can conclude and get the China data for GBP Table 3.39.

As shown above Table 3.40, there is no remarkable difference in GBP in port of Shanghai, port of Hong Kong, and port of Shenzhen through the result of data analysis. And also, the berth lengths are almost 350 m, the dwell times are similar except Cosco-HIT, and the container yard area, number of quay crane, yard crane and yard tractor also not have remarkable difference. On another hand, in every Chinese port has not enough data to make a regress analysis. Therefore, we can put Chinese ports into one basket together; there is no necessity to make a distribution analysis in port of China.

Chapter 4
Result of Analysis

4.1 Analysis Result of Port of Korea

4.1.1 GBP Port of Busan

As already discussed in Chap. 3, independent variables are composed of dwell time, yard area, number of yard crane, berth length, number of yard tractor, and quay crane. When backward elimination method is applied to the model, the R square which means the coefficient of determination (SSR/SST) ranges from 0.931 to 0.961. These six independent factors explain model in a good fitness.

In ANOVA analysis, the significant probability is p = 0.113 in applying six independent variables and reaches 0.00 if the independent variable with high significant value removed orderly. If we set the level of significance 5 %, the regression model is valid only when yard area as independent factor removed. However if we check significance probability on each independent variables, number of yard tractor and length of berth can be accepted as valid independent factor of which value are 0.006 and 0.028 (Tables 4.1, 4.2 and 4.3).

In coefficients table, the Standardized Coefficients Beta value appear the minus value, such as yard area, yard crane, and dwell time, this means that as the mount of GBP increasing, the quantity of yard crane, yard area, and dwell time should be reduced. So it is strange results. The reason is that before we make a liner analysis, in the collected data, each independent of each terminal and also the dependent data of each terminal has remarkable difference of each terminal, for example, for two terminals, the GBP of first terminal is less than the second one, however, the mount of yard crane, yard area, and dwell time are more than the first one. So the small quantity of dependent variable and the big quantity of independent variable cause to

B. Lu and S. Wang, *Critical Factors for Berth Productivity in Container Terminal*,
DOI 10.1007/978-981-10-2431-3_4

Table 4.1 Korea data regression model summary[e]

Model	R	R square	Adjusted R square	Std. error of the estimate
1	0.980[a]	0.961	0.843	14.70066
2	0.980[b]	0.961	0.895	12.02201
3	0.980[c]	0.960	0.920	10.48764
4	0.970[d]	0.941	0.905	11.43865
5	0.965[e]	0.931	0.908	11.28533

[a]Predictors: (Constant), Dwell Time, Yard Area, Yard Crane, Berth Length, Yard_Tractor, Quay_Crane
[b]Predictors: (Constant), Dwell_Time, Yard_Crane, Berth_Length, Yard_Tractor, Quay_Crane
[c]Predictors: (Constant), Yard_Crane, Berth_Length, Yard_Tractor, Quay_Crane
[d]Predictors: (Constant), Berth_Length, Yard_Tractor, Quay_Crane
[e]Predictors: (Constant), Berth_Length, Yard_Tractor

Table 4.2 Korea data ANOVA[f] analysis summary

Model		Sum of squares	df	Mean square	F	Sig.
1	Regression	10582.855	6	1763.809	8.162	0.113[a]
	Residual	432.219	2	216.109		
	Total	11015.074	8			
2	Regression	10581.487	5	2116.297	14.643	0.026[b]
	Residual	433.586	3	144.529		
	Total	11015.074	8			
3	Regression	10575.112	4	2643.778	24.036	0.005[c]
	Residual	439.962	4	109.991		
	Total	11015.074	8			
4	Regression	10360.860	3	3453.620	26.395	0.002[d]
	Residual	654.214	5	130.843		
	Total	11015.074	8			
5	Regression	10250.922	2	5125.461	40.244	0.000[e]
	Residual	764.152	6	127.359		
	Total	11015.074	8			

[a]Predictors: (Constant), Dwell_Time, Yard_Area, Yard_Crane, Berth_Length, Yard_Tractor, Quay_Crane
[b]Predictors: (Constant), Dwell_Time, Yard_Crane, Berth_Length, Yard_Tractor, Quay_Crane
[c]Predictors: (Constant), Yard_Crane, Berth_Length, Yard_Tractor, Quay_Crane
[d]Predictors: (Constant), Berth_Length, Yard_Tractor, Quay_Crane
[e]Predictors: (Constant), Berth_Length, Yard_Tractor
[f]Dependent Variable: GBP

the Standardized Coefficients Beta value appear the minus value. And also, the independent variables which for dependent variable GBP have least influence have been eliminate firstly.

Table 4.3 Korea data coefficients[a]

Model		Unstandardized coefficients		Standardized coefficients	t	Sig.
		B	Std. error	Beta		
1	(Constant)	−151.431	71.888		−2.106	0.170
	Length Berth	0.449	0.261	0.402	1.720	0.227
	Yard Area	−8.37E−006	0.000	−0.012	−0.080	0.944
	Quay Crane	22.417	29.020	0.298	0.772	0.521
	Yard Crane	−2.374	2.781	−0.262	−0.854	0.483
	Yard Tractor	4.866	2.154	0.594	2.259	0.152
	Dwell time	−0.285	1.707	−0.030	−0.167	0.883
2	(Constant)	−152.24	58.199		−2.616	0.079
	Length Berth	0.445	0.210	0.399	2.118	0.124
	Quay Crane	22.820	23.369	0.303	0.977	0.401
	Yard Crane	−2.368	2.273	−0.261	−1.042	0.374
	Yard Tractor	4.840	1.742	0.591	2.778	0.069
	Dwell time	−0.293	1.394	−0.031	−0.210	0.847
3	(Constant)	−157.83	45.137		−3.497	0.025
	Length Berth	0.435	0.178	0.389	2.443	0.071
	Quay Crane	25.610	16.772	0.340	1.527	0.201
	Quay Crane	−2.553	1.829	−0.282	−1.396	0.235
	Yard Tractor	4.856	1.519	0.593	3.198	0.033
4	(Constant)	−127.266	43.045		−2.957	0.032
	Length Berth	0.391	0.191	0.350	2.048	0.096
	Quay Crane	14.916	16.272	0.198	0.917	0.401
	Yard Tractor	4.190	1.572	0.511	2.665	0.045
5	(Constant)	−118.936	41.510		−2.865	0.029
	Length Berth	0.476	0.165	0.426	2.883	0.028
	Yard Tractor	5.092	1.211	0.621	4.206	0.006

[a]Dependent Variable: GBP

The regression formulate which is composed of selected variables is shown in Formulation-2:

$$\text{GBP} = -118.936 + 0.476X1 + 5.092X2 \qquad \text{(Formulation-2)}$$

where, *X1* = *Length of Berth*
X2 = *Quantities of Yard Tractor*

4.1.2 GBP—Port of Gwangyang

When enter method is applied to the model, even though the R square is 1.0, mean square error is 0. This means any significant result can't be extracted notwithstanding applying all methods. This is caused by small number of sample, long-term storage of container in terminal.

4.2 Analysis Result of Port of China

4.2.1 GBP—Port of China

As already discussed in Analysis result of port of China, independent variables are composed of number of yard tractor, berth of length, the yard area, dwell time, number of terminal crane, number of quay crane. When backward elimination method is applied to the model, the R square which means the coefficient of determination (SSR/SST) is from 0.510 to 0.638. The independent factor is lower than port of Busan, yet the extent is enough to explain model in a good fitness.

In ANOVA analysis, the significant probability is $p = 0.740$ in applying six independent variables and reaches 0.031 if the independent variable with high significant value removed orderly. If we set the level of significance 10 %, the regression model is valid only when yard tractor, as independent factor removed. And also if we check significance probability on each independent variable, Yard Area can be accepted as valid independent factor of which value is 0.031 (Tables 4.4, 4.5 and 4.6).

In China coefficients analysis, just like the Korea coefficients analysis, the Standardized Coefficients Beta values appear the minus value, such as quay crane and yard tractor, also the small quantity of dependent variable and the big quantity of independent variable cause to the Standardized Coefficients Beta value appear the minus value. And also, the independent variables which for dependent variable GBP have least influence have been eliminate firstly.

The regression formulate which is composed of selected variables is shown in Formulation-3:

$$\mathrm{GBP} = 92.240 + 0.002X1 \qquad \text{(Formulation-3)}$$

where, $X1$ = *Yard Area*

Table 4.4 China data regression model summary[e]

Model	R	R square	Adjusted R square	Std. error of the estimate
1	0.799[a]	0.638	−0.448	63.17231
2	0.799[b]	0.638	0.035	51.58009
3	0.797[c]	0.635	0.271	44.83675
4	0.781[d]	0.610	0.375	41.48632
5	0.758[e]	0.574	0.432	39.56359
6	0.714[f]	0.510	0.440	39.29048

[a]Predictors: (Constant), Dwell_Time, Yard_Tractor, Quay_Crane, Yard_Area, Yard_Crane, Berth_Length
[b]Predictors: (Constant), Dwell_Time, Quay_Crane, Yard_Area, Yard_Crane, Berth_Length
[c]Predictors: (Constant), Dwell_Time, Yard_Area, Yard_Crane, Berth_Length
[d]Predictors: (Constant), Dwell_Time, Yard_Area, Berth_Length
[e]Predictors: (Constant), Yard_Area, Berth_Length
[f]Predictors: (Constant), Yard_Area

Table 4.5 China data ANOVA[f]

Model		Sum of squares	df	Mean square	F	Sig.
1	Regression	14065.226	6	2344.204	0.587	0.740[a]
	Residual	7981.480	2	3990.740		
	Total	22046.707	8			
2	Regression	14065.190	5	2813.038	0.057	0.515[b]
	Residual	7981.516	3	2660.505		
	Total	22046.707	8			
3	Regression	14005.370	4	3501.343	1.742	0.302[c]
	Residual	8041.336	4	2010.334		
	Total	22046.707	8			
4	Regression	13441.132	3	4480.377	2.603	0.164[d]
	Residual	8605.574	5	1721.115		
	Total	22046.707	8			
5	Regression	12655.042	2	6327.521	4.042	0.077[e]
	Residual	9391.665	6	1565.277		
	Total	22046.707	8			
6	Regression	11240.513	1	11240.513	7.281	0.031[f]
	Residual	10806.193	7	1543.742		
	Total	22046.707	8			

[a]Predictors: (Constant), Dwell_Time, Yard_Tractor, Quay_Crane, Yard_Area, Yard_Crane, Berth_Length
[b]Predictors: (Constant), Dwell_Time, Quay_Crane, Yard_Area, Yard_Crane, Berth_Length
[c]Predictors: (Constant), Dwell_Time, Yard_Area, Yard_Crane, Berth_Length
[d]Predictors: (Constant), Dwell_Time, Yard_Area, Berth_Length
[e]Predictors: (Constant), Yard_Area, Berth_Length
[f]Predictors: (Constant), Yard_Area

Table 4.6 China data coefficients[a]

Model		Unstandardized coefficients		Standardized coefficients	t	Sig.
		B	Std. error	Beta		
1	(Constant)	−989.437	1487.764		−0.665	0.574
	Berth Length	2.964	4.015	0.905	0.738	0.537
	Yard Area	0.003	0.002	0.914	1.431	0.289
	Quay Crane	−2.509	24.278	−0.054	−0.103	0.927
	Yard Crane	2.268	8.475	0.186	0.268	0.814
	Yard Tractor	−0.018	5.908	−0.002	−0.003	0.998
	Dwell_Time	37.666	74.348	0.567	0.507	0.663
2	(Constant)	−988.128	1161.206		−0.851	0.457
	Berth Length	2.961	3.154	0.904	0.939	0.417
	Yard Area	0.003	0.001	0.913	2.173	0.118
	Quay Crane	−2.546	16.980	−0.055	−0.150	0.890
	Yard Crane	2.249	4.729	0.184	0.476	0.667
	Dwell_Time	37.624	59.664	0.567	0.631	0.573
3	(Constant)	−996.231	1008.301		−0.988	0.379
	Berth_Length	2.955	2.741	0.902	1.078	0.342
	Yard Area	0.003	0.001	0.904	2.500	0.067
	Yard_Crane	2.161	4.078	0.177	0.530	0.624
	Dwell_Time	38.796	51.417	0.584	0.755	0.493
4	(Constant)	−766.174	842.000		−0.910	0.405
	Berth_Length	2.375	2.325	0.725	1.021	0.354
	Yard_Area	0.003	0.001	0.880	2.651	0.045
	Dwell_Time	30.699	45.424	0.462	0.676	0.529
5	(Constant)	−263.262	375.703		−0.701	0.510
	Berth_Length	0.003	0.001	1.026	4.103	0.006
	Yard_Area	0.003	0.001	0.877	2.769	0.032
6	(Constant)	92.240	35.835		2.574	0.037
	Yard_Area	0.002	0.001	0.714	2.698	0.031

[a]Dependent Variable: GBP

4.3 Compare the Analysis Result of Port Between China and Korea Port

Because there are many difference between Korea and China ports, for instance the geography location, the different yard area, the different dwell time, the difference of facilities, the difference of equipments, information technology, the method of management, throughputs, and container disposal capability. So the productivity also has otherness, and the factors which related to the productivity and the sequence of the factors have otherness, too.

Data of Port of China and Korea was undertaken to make an analysis. The remarkable difference between in China and in Korea port is detected through the result of the analysis. In China case, the highest and direct scope of factors which relativity of container terminal productivity are all about container facilities such as Yard Area and Berth Length. It is because that in china, the area of Chinese ports yard are much bigger than the area of Korean ports, and the port facilities such as yard area and berth length.

In Korea case, the highest and direct scope of factors which relativity of container terminal productivity is yard tractor and berth length. From the results of Korean ports data regression analysis, we can see that, because of the area of Korean ports are smaller than Chinese ports, the method of Korean ports to enhance berth productivity are reducing the yard tractor operation time from the berth to container yard and improving the speed of yard tractor and other container equipments such as yard crane and quay crane; also because the length of Chinese ports berth are almost 335–350 m from the data, otherwise, the berth length of Korean ports have various types such as 120, 200, 240, 300, 350. So the different berth lengths affect berth productivity of Korean ports, owing to the capability of equipped quay crane for the long berth, and the quantity are better than short berth.

Chapter 5
Conclusion

This study aims at extracting the major factors which affect gross berth productivity (GBP), and then identify the relationships between dependent variable and independent variable and the order of importance among the factors. Furthermore the paper tries to compare the difference between China and Korea mega-container terminals as to give the guideline to terminal operators and to improve berth productivity.

For this purpose, we extract potential factors from literature, interview and site observation. After verifying potential factors, I made the regression model and did model test using ANOVA statistics. The experimental data is collected in scope of container terminal which opened in China and South Korea.

The remarkable difference between in China and in Korea is detected through the result of the analysis. In Korea case, the highest and direct scope of factors which relativity of container terminal productivity is yard tractor and berth length. It is because that the area of Korean ports are smaller than Chinese ports, the method of Korean ports to enhance berth productivity are to increase number of the yard tractor, then reduce cycle time from the berth to container yard and improve the speed of yard tractor and other container equipments such as yard crane and quay crane; also because the length of Chinese ports berth are almost 335–350 m from the data, otherwise, the berth length of Korean ports have various types such as 120, 200, 240, 300, 350. So the different berth lengths affect berth productivity of Korean ports.

From a standpoint of shipping company being customer of container terminal, the company use port of Korea Busan in expecting of speedy loading, unloading and trans-shipment, and use the port of Korea Gwangyang in expecting of long-term stock as like depot.

In China case, the highest and direct scope of factors which relativity of container terminal productivity are all about container facilities such as Yard Area and Berth Length. It is because that in china, the area of Chinese ports yard are much bigger than the area of Korean ports. In comparativeness, Chinese ports take the advantage of the predominance to enhance the berth productivity.

B. Lu and S. Wang, *Critical Factors for Berth Productivity in Container Terminal*,
DOI 10.1007/978-981-10-2431-3_5

The results also explain that for Chinese terminal, the more important function of China is container storage, turnover and taking the advantage of port facilities to improve the container disposal quantities. These results also suggest the major function of port of China for improving is container transfer and storage in container terminal.

In short, this study reveals that the improvement of GBP in container terminal has closely relationship with factors such as number of quay crane, number of yard tractor, length of berth, the yard area, dwell time, and number of yard tractors. The comparative result of this study will contribute to give tips for the priority of investment for an improvement of productivity in constructing or under operation.

But this study has the many critical points. Firstly, the synergistic effect which follows in scale can be excluded because resources are divided arithmetically by number of berth in GBP. Recently, the fact is reported that the performance of crane, especially twin-lift, above 224 bay Quay Crane or automatic level affect improvement of productivity. But in this study the technology factors can not be considered due to measurement. And we have to point that there is a difficulty to generalize owing to smaller sample and restricted area.

So, in the future analysis, I shall make an analysis about why there are difference between the Busan and Gwangyang which have the similar facilities and equipment however have so many differences about the throughput, Berth Occupancy and GBP; and also shall go on collecting the China and Korea port data to perfect the analysis model.

Bibliography

Accorsi R, Manzini R, Ferrari E (2014) A comparison of shipping containers from technical, economic and environmental perspectives. Transp Res Part D Transp Environ 26:52–59

Acosta M, Coronado D, Mar Cerban M (2007) Port Competitiveness in Container Traffic from an Internal Point of View: the Experience of the Port of Algeciras Bay. Marit Policy Manage 34(5):501–520

Aerts G, Grage T, Dooms M, Haezendonck E (2014) Public-private partnerships for the provision of port infrastructure: an explorative multi-actor perspective on critical success factors1. Asian J Shipping Logistics 30(3):273–298

Alfa MJ, Lo E, Olson N, MacRae M, Buelow-Smith L (2015) Use of a daily disinfectant cleaner instead of a daily cleaner reduced hospital-acquired infection rates. Am J Infect Control 43 (2):141–146

Ambrosino D, Sciomachen A (2014) Location of mid-range dry ports in multimodal logistic networks. Proc Soc Behav Sci 108(8):118–128

Arıkan E, Fichtinger J, Ries JM (2014) Impact of transportation lead-time variability on the economic and environmental performance of inventory systems. Int J Prod Econ 157:279–288

Assadipour G, Ke GY, Verma M (2015) Planning and managing intermodal transportation of hazardous materials with capacity selection and congestion. Transp Res Part E Logistics Transp Rev 76:45–57

Bandara YM, Garaniya V, Chin C, Leong ZH (2015) Improving logistics management using foldable/collapsible containers: a case study. Asian J Shipping Logistics 31(1):161–185

Blas J (2015) Chapter 33—Stress in birds. In: Scanes CG (ed) Sturkie's Avian Physiology, 6th edn. Academic Press, San Diego, pp 769–810

Bouchery Y, Fransoo J (2014) Cost, carbon emissions and modal shift in intermodal network design decisions. Int J Prod Econ (Available online 9 December 2014)

Boysen Nils, Emde Simon, Hoeck Michael, Kauderer Markus (2015) Part logistics in the automotive industry: decision problems, literature review and research agenda. Eur J Oper Res 242(1):107–120

Bray S, Caggiani L, Dell'Orco M, Ottomanelli M (2014) Features selection based on fuzzy entropy for Data Envelopment Analysis applied to transport systems. Transp Res Proc 3:602–610

Bray S, Caggiani L, Ottomanelli M (2015) Measuring transport systems efficiency under uncertainty by fuzzy sets theory based Data Envelopment Analysis: theoretical and practical comparison with traditional DEA model. Transp Res Proc 5:186–200

Brown KL, Wray S (2014) 7—Control of airborne contamination in food processing. In: Lelieveld HLM, Holah JT, Napper D (eds) Hygiene in food processing, 2nd edn. Woodhead Publishing, pp 174–202

Burke B, Stewart CL (2014) Chapter One—Functional architecture of the cell's nucleus in development, aging, and disease. In: Stewart CL (ed) Current topics in developmental biology, vol 109. Academic Press, pp 1–52

B. Lu and S. Wang, *Critical Factors for Berth Productivity in Container Terminal*,
DOI 10.1007/978-981-10-2431-3

Cabrera M, Suárez-Alemán A, Trujillo L (2015) Public-private partnerships in Spanish Ports: current status and future prospects. Utilities Policy 32:1–11

Cai Y, Zhang YW (2006) Simulation optimization for berth allocation in container terminals. Chin J Constr Mach 4(2):228–232

Caris A, Limbourg S, Macharis C, van Lier T, Cools M (2014) Integration of inland waterway transport in the intermodal supply chain: a taxonomy of research challenges. J Trans Geogr 41:126–136

Carlo HJ, Vis IFA, Roodbergen KJ (2014) Storage yard operations in container terminals: literature overview, trends, and research directions. Eur J Oper Res 235(2):412–430

Castillo-Manzano JI, Castro-Nuño M, Fageda X, Gonzalez-Aregall M (2015) Evaluating the effects of the latest change in Spanish port legislation: Another "turn of the screw" in port reform? Case Studies on Transport Policy (Available online 19 February 2015)

CCDoTT (2010) Center for the commercial deployment of transportation technologies. May 2010

Chang V, Tovar B (2014a) Efficiency and productivity changes for Peruvian and Chilean ports terminals: a parametric distance functions approach. Transp Policy 31:83–94

Chang V, Tovar B (2014b) Drivers explaining the inefficiency of Peruvian and Chilean ports terminals. Transp Res Part E Logistics Transp Rev 67:190–203

Chang C-H, Xu J, Song D-P (2014) An analysis of safety and security risks in container shipping operations: a case study of Taiwan. Saf Sci 63:168–178

Chapter 3—Natural Gas liquefaction. In: Mokhatab S, Mak JY, Valappil JV, Wood DA (eds) Handbook of liquefied natural gas. Gulf Professional Publishing, Boston, pp 147–183

Chapter 7—LNG plant and regasification terminal operations. In: Mokhatab S, Mak JY, Valappil JV, Wood DA (eds) Handbook of liquefied natural gas. Gulf Professional Publishing, Boston, pp 297–320

Chen T (1998) Land utilisation of the container terminals: a global perspective. Marit Policy Manage 25(4):289–303

Chen L, Wang N, Sun D, Li L (2014) Microwave-assisted acid hydrolysis of proteins combined with peptide fractionation and mass spectrometry analysis for characterizing protein terminal sequences. J Proteomics 100(4):68–78

Chiu R-H, Yen DC (2015) Application of organizational life cycle theory for port reform initiatives in Taiwan. Res Transp Bus Manag 14:14–24

Chou CC (2002) Analysis of container throughput of major ports in Far Eastern region. Marit Res J (12):59–71

Cho H (2014) Determinants and effects of logistics costs in container ports: the transaction cost economics perspective. Asian J Shipping Logistics 30(2):193–215

Clott C, Hartman BC, Ogard E, Gatto A (2015) Container repositioning and agricultural commodities: shipping soybeans by container from US hinterland to overseas markets. Res Transp Bus Manag 14:56–65

Cullinanea K, Wang TF, Song DW, Ji P (2006) The technical efficiency of container ports: comparing data envelopment analysis and stochastic frontier analysis. Transport Res Part A: Policy Pract 40(4):354–374

De P (2006) Total Factor Productivity Growth: Indian Ports in the Era of Globalization. Marit Econ Logistics 8(4):366–386

de la O G, Matis T (2014) Supply Chain Strategies for the international interoceanic Mazatlan-Houston Logistic Corridor. J Appl Res Technol 12(4):666–673

de Almeida Guimarães V, Junior ICL, de Almada PA (2014) Garcia, XI Congreso De Ingeniería Del Transporte (CIT 2014) Environmental performance of Brazilian container terminals: a data envelopment analysis approach. Procedia Soc Behav Sci 160:178–187

Derakhshani SM, Schott DL, Lodewijks G (2015) Micro–macro properties of quartz sand: Experimental investigation and DEM simulation. Powder Technol 269:127–138

Dragović B, Park NK, Radmilović Z, Maraš V (2005) Simulation modelling of ship-berth link with priority service. Marit Econ Logistics 7(4):316–335

Dulebenets MA, Golias MM, Mishra S, Heaslet WC (2015) Evaluation of the floaterm concept at marine container terminals via simulation. Simul Model Pract Theory 54:19–35

Fatimah UZAU, Strohbehn CH, Arendt SW (2014) An empirical investigation of food safety culture in onsite foodservice operations. Food Control 46:255–263

Featherstone S (2015) 12—Sterilisation systems. In: Featherstone S (ed) Woodhead publishing series in food science, technology and nutrition. Woodhead Publishing, pp 239–267

Filha MHNLS, Berry C, Regis L (2014) Chapter Three—Lysinibacillus sphaericus: toxins and mode of action, applications for mosquito control and resistance management. In: Dhadialla TS, Gill SS (eds) Advances in insect physiology, vol 47. Academic Press, pp 89–176

Fu Y-M, Diabat A (2015) A Lagrangian relaxation approach for solving the integrated quay crane assignment and scheduling problem. Appl Math Model 39(3–4):1194–1201

Gadipelli S, Guo ZX (2015) Graphene-based materials: synthesis and gas sorption, storage and separation. Prog Mater Sci 69:1–60

Govindan K, Diabat A, Shankar KM (2014) Analyzing the drivers of green manufacturing with fuzzy approach. J Cleaner Prod (Available online 19 March 2014)

Ghaderi H, Cahoon S, Nguyen H-O (2015) An investigation into the non-bulk rail freight transport in Australia. Asian J Shipping Logistics 31(1):59–83

GU M (2006) Energetic Singapore container terminal. Port Econ (2):60–60

Güner S (2015) Investigating infrastructure, superstructure, operating and financial efficiency in the management of Turkish seaports using data envelopment analysis. Transp Policy 40:36–48

Handelsman DJ (2016) Chapter 24—Performance enhancing hormones in sports doping. In: Larry Jameson J, De Groot LJ, de Kretser DM, Giudice LC, Grossman AB, Melmed S, Potts JT, Weir GC, Saunders WB (eds) Endocrinology: adult and pediatric, 7th edn, Philadelphia, pp 441–454

He J, Huang Y, Yan W (2015) Yard crane scheduling in a container terminal for the trade-off between efficiency and energy consumption. Adv Eng Inform 29(1):59–75

He J, Huang Y, Yan W, Wang S (2015) Integrated internal truck, yard crane and quay crane scheduling in a container terminal considering energy consumption. Exp Syst Appl 42 (5):2464–2487

Hirdaris SE, Bai W, Dessi D, Ergin A, Gu X, Hermundstad OA, Huijsmans R, Iijima K, Nielsen UD, Parunov J, Fonseca N, Papanikolaou A, Argyriadis K, Incecik A (2014) Loads for use in the design of ships and offshore structures. Ocean Eng 78(1):131–174

Homayouni SM, Tang SH, Motlagh O (2014) A genetic algorithm for optimization of integrated scheduling of cranes, vehicles, and storage platforms at automated container terminals. J Comput Appl Math 270:545–556

Jeevan J, Chen S, Lee E (2015) The challenges of Malaysian dry ports development. Asian J Shipping Logistics 31(1):109–134

Jiang J, Lee LH, Chew EP, Gan CC (2015) Port connectivity study: an analysis framework from a global container liner shipping network perspective. Transp Res Part E Logistics Transp Rev 73:47–64

Kim KY, Kim KH (1999) A routing algorithm for a single straddle carrier to load export cotainers onto a containership. Int J Prod Econ (59):425–433

Kim S, Bedigrew K, Guda T, Maloney WJ, Park S, Wenke JC, Yang YP (2014) Novel osteoinductive photo-cross-linkable chitosan-lactide-fibrinogen hydrogels enhance bone regeneration in critical size segmental bone defects. Acta Biomater 10(12):5021–5033

Korhonen P, Kaila J (2015) Waste container weighing data processing to create reliable information of household waste generation. Waste Manag (Available online 9 March 2015)

Lam JSL, Lai K (2014) Developing environmental sustainability by ANP-QFD approach: the case of shipping operations. J Cleaner Prod (Available online 30 September 2014)

Lau Y, Tam K, Ng AKY, Pallis AA (2014) Cruise terminals site selection process: an institutional analysis of the Kai Tak Cruise Terminal in Hong Kong. Res Transp Bus Manag 13:16–23

Le-Griffin HD, Murphy M (2007) Container terminal productivity: Experiences at the ports of Los Angeles and Long Beach. Proceeding of NUF Conference: 1–21, Nov 2007

Leach PT, Kenney JW, Connor DA, Gould TJ (2015) Thyroid receptor β involvement in the effects of acute nicotine on hippocampus-dependent memory. Neuropharmacology 93:155–163

Lee CB, Wan J, Shi W, Li K (2014) A cross-country study of competitiveness of the shipping industry. Transp Policy 35:366–376

Lee T, Yeo G-T, Thai VV (2014) Structural analysis of port brand equity using structural equation modeling1. Asian J Shipping Logistics 30(3):349–372

Li L, Wang B, Cook DP (2014) Enhancing green supply chain initiatives via empty container reuse. Transp Res Part E Logistics Transp Rev 70:190–204

Li L, Wang B, Cook DP (2015) Reprint of "Enhancing green supply chain initiatives via empty container reuse". Transp Res Part E Logistics Transp Rev 74:109–123

Liu B, Qiu G, Liu C (2003) Linear regression model of China's container port throughput. J Dalian Marit Univ 29(2):27–30

Liang C, Gu T, Lu B, Ding Y (2015) Genetic mechanism-based coupling algorithm for solving coordinated scheduling problems of yard systems in container terminals. Comput Ind Eng (Available online 14 March 2015)

Lin D-Y, Lee Y-J, Lee Y (2015) The container retrieval problem with respect to relocation. Transp Res Part C Emerg Technol 52:132–143

Loh HS, Van Thai V (2014) Managing port-related supply chain disruptions: a conceptual paper. Asian J Shipping Logistics 30(1):97–116

Lee C-Y, Lee HL, Zhang J (2015) The impact of slow ocean steaming on delivery reliability and fuel consumption. Transp Res Part E Logistics Transp Rev 76:176–190

Mabrouki C, Bentaleb F, Mousrij A (2014) A decision support methodology for risk management within a port terminal. Saf Sci 63:124–132

Macário R, Ribeiro J, Costa J (2015) Understanding pitfalls in the application of PPPs in transport infrastructure in Portugal. Transp Policy (Available online 3 April 2015)

Mansouri SA, Lee H, Aluko O (2015) Multi-objective decision support to enhance environmental sustainability in maritime shipping: a review and future directions. Transp Res Part E Logistics Transp Rev (Available online 21 February 2015)

Massei M, Poggi S, Agresta M, Ferrando A (2014) Development planning based on interoperable agent driven simulation. J Comput Sci 5(3):395–407

Mathisen TA, Hanssen T-ES (2014) The academic literature on intermodal freight transport. Transp Res Proc 3:611–620

Michele A, Patrizia S (2014) Strategic determinants of terminal operating system choice: an empirical approach using multinomial analysis. Transp Res Proc 3:592–601

Mili K (2014) Six Sigma Approach for the Straddle carrier routing problem. Proc Soc Behav Sci 111(5):1195–1205

Monios J, Bergqvist R (2015) Intermodal terminal concessions: lessons from the port sector. Res Transp Bus Manag 14:90–96

Ng AKY, Ducruet C, Jacobs W, Monios J, Notteboom T, Rodrigue J-P, Slack B, Tam K, Wilmsmeier G (2014) Port geography at the crossroads with human geography: between flows and spaces. J Transp Geogr 41:84–96

Notteboom TE (2004) Container shipping and ports: an overview. Review of Network Economics 3(2):1–21

Ouyang M (2014) Review on modeling and simulation of interdependent critical infrastructure systems. Reliab Eng Syst Saf 121:43–60

Pallme D, Lambert B, Miller C, Lipinski M (2015) A review of public and private intermodal railroad development in the Memphis region. Res Transp Bus Manag 14:44–55

Panayides PM, Parola F, Lam JSL (2015) The effect of institutional factors on public–private partnership success in ports. Transp Res Part A Policy Pract 71:110–127

Pant R, Barker K, Ramirez-Marquez JE, Rocco CM (2014) Stochastic measures of resilience and their application to container terminals. Comput Ind Eng 70:183–194

Park NK, Lee SW, Park HR, Kwon HK (2004) A Simulation model for improving the productivity of container handling in the container terminal. J Navig Port Res 28(4):285–291

Peng Fei Z, Zi Jian G, Xiang Qun S (2006) Simulation study on container terminal performance. International Conference on Management Science 177–182

Pestana C, Peypoch N (2007) Comparing productivity change in Italian and Portuguese seaports using the Luenberger indicator approach. Marit Econ Logistics 9(2):138–147

Rabbani A, Zamani M, Yazdani-Chamzini A, Zavadskas EK (2014) Proposing a new integrated model based on sustainability balanced scorecard (SBSC) and MCDM approaches by using linguistic variables for the performance evaluation of oil producing companies. Exp Syst Appl 41(16):7316–7327

Rechard RP, Lee JH, Hardin EL, Bryan CR (2014) Waste package degradation from thermal and chemical processes in performance assessments for the Yucca Mountain disposal system for spent nuclear fuel and high-level radioactive waste. Reliab Eng Syst Saf 122:145–164

Robenek T, Umang N, Bierlaire M, Ropke S (2014) A branch-and-price algorithm to solve the integrated berth allocation and yard assignment problem in bulk ports. Eur J Oper Res 235 (2):399–411

Rodrigues VS, Beresford A, Pettit S, Bhattacharya S, Harris I (2014) Assessing the cost and CO_2e impacts of rerouteing UK import containers. Transp Res Part A Policy Pract 61:53–67

Sachish A (1996) Productivity functions as a managerial tool in Israeli ports. Marit Policy Manage 23(23):341–369

Santos BF, Limbourg S, Carreira JS (2015) The impact of transport policies on railroad intermodal freight competitiveness—the case of Belgium. Transp Res Part D Transp Environ 34:230–244

Satta G, Parola F, Persico L (2014) Temporal and Spatial Constructs in Service Firms' internationalization patterns: the determinants of the accelerated growth of emerging MNEs. J Int Manag 20(4):421–435

Sayareh J, Alizmini HR (2014) A hybrid decision-making model for selecting container seaport in the Persian Gulf. Asian J Shipping Logistics 30(1):75–95

Schweizer D, Serno T, Goepferich A (2014) Controlled release of therapeutic antibody formats. Eur J Pharm Biopharm 88(2):291–309

Shafieezadeh A, Burden LI (2014) Scenario-based resilience assessment framework for critical infrastructure systems: case study for seismic resilience of seaports. Reliab Eng Syst Saf 132:207–219

Sheng PC (2007) Comparing China Container Terminal with Foreign Mega-Container Terminal Productivity. Transport market (3):74–77

Shih DT-B, Burnouf T (2015) Preparation, quality criteria, and properties of human blood platelet lysate supplements for ex vivo stem cell expansion. New Biotechnol 32(1):199–211

Simcik M, Ruzicka MC (2015) CFD model for pneumatic mixing with bubble chains: application to glass melts. Chem Eng Sci 127(4):344–361

Smith-Pardo JP, Reyes JC, Ardila-Bothia L, Villamizar-Gonzalez JN, Ardila-Giraldo OA (2015) Effect of live load on the seismic design of single-story storage structures under unidirectional horizontal ground motions. Eng Struct 93:50–60

Song DW, Cullinane K (2006) Estimating the Relative Efficiency of European Container Ports: A Stochastic Frontier Analysis. Research in Transportation Economics 16(1):85–115

Song B, Cui Y (2014) Productivity changes in Chinese Container Terminals 2006–2011. Transp Policy 35:377–384

SteadieSeifi M, Dellaert NP, Nuijten W, Van Woensel T, Raoufi R (2014) Multimodal freight transportation planning: a literature review. Eur J Oper Res 233(1):1–15

Sutrisnowati RA, Bae H, Song M (2014) Bayesian network construction from event log for lateness analysis in port logistics. Comput Ind Eng (Available online 21 November 2014)

Talley WK, Ng MW, Marsillac E (2014) Port service chains and port performance evaluation. Transp Res Part E Logistics Transp Rev 69:236–247

Tan Z, Li W, Zhang X, Yang H (2015) Service charge and capacity selection of an inland river port with location-dependent shipping cost and service congestion. Transp Res Part E Logistics Transp Rev 76:13–33

Tierney K, Voß S, Stahlbock R (2014) A mathematical model of inter-terminal transportation. Eur J Oper Res 235(2):448–460

Ting C-J, Wu K-C, Chou H (2014) Particle swarm optimization algorithm for the berth allocation problem. Exp Syst Appl 41(4), Part 1:1543–1550

Thorp JH, Covich AP (2015) Chapter 2–Overview of inland water habitats. In: Thorp JH, Rogers DC (eds) Thorp and Covich's freshwater invertebrates, 4th edn. Academic Press, Boston, pp 23–56

Tovar B, Wall A (2015) Can ports increase traffic while reducing inputs? Technical efficiency of Spanish Port Authorities using a directional distance function approach. Transp Res Part A Policy Pract 71 (January 2015)

Tran NK, Haasis H-D (2015) An empirical study of fleet expansion and growth of ship size in container liner shipping. Int J Prod Econ 159:241–253

Tsai C (2014) The Organizational Climate and Employees' Job Satisfaction in the Terminal Operation Context of Kaohsiung Port1. Asian J Shipping Logistics 30(3):373–392

Ursavas E (2014) A decision support system for quayside operations in a container terminal. Decis Support Syst 59:312–324

Venugopalan J, Sarath VS, Pillai RJ, Anantha Krishnan S, Anbuudayasankar SP (2014) Analysis of decision models in supply chain management. Proc Eng 97

Wagner R (2009) Secondary batteries—lead—acid systems|flooded batteries. In: Garche J (ed) Encyclopedia of electrochemical power sources. Elsevier, Amsterdam, pp 677–692

Wang Y, Cullinane K (2014) Traffic consolidation in East Asian container ports: a network flow analysis. Transp Res Part A Policy Pract 61:152–163

Wang N, Jin B, Lim A (2015) Target-guided algorithms for the container pre-marshalling problem. Omega 53:67–77

Wang S, Liu Z, Bell MGH (2015) Profit-based maritime container assignment models for liner shipping networks. Transp Res Part B Methodol 72:59–76

Wang X, Liu Z, Shang Y (2014) Sustainable analysis architecture and optimization of urban freight under congestion. Proc Soc Behav Sci 138(14):418–426

Wang GWY, Pallis AA (2014) Incentive approaches to overcome moral hazard in port concession agreements. Transp Res Part E Logistics Transp Rev 67:162–174

Wang GWY, Pallis AA, Notteboom TE (2014) Incentives in cruise terminal concession contracts. Res Transp Bus Manag 13:36–42

Wanke P, Barros CP, Figueiredo Otavio (2014) Measuring efficiency improvement in Brazilian trucking: a distance friction minimization approach with fixed factors. Measurement 54:166–177

Wilmsmeier G, Monios J (2015) Institutional structure and agency in the governance of spatial diversification of port system evolution in Latin America. J Transp Geogr (Available online 4 March 2015)

Wilmsmeier G, Monios J, Pérez-Salas G (2014) Port system evolution—the case of Latin America and the Caribbean. J Transp Geogr 39:208–221

Wilmsmeier G, Monios J, Rodrigue J-P (2015) Drivers for Outside-In port hinterland integration in Latin America: the case of Veracruz, Mexico. Res Transp Bus Manag 14:34–43

Wong EYC, Tai AH, Lau HYK, Raman M (2015) An utility-based decision support sustainability model in slow steaming maritime operations. Transp Res Part E Logistics Transp Rev (Available online 20 March 2015)

Wu WM, Lin J-R (2015) Productivity growth, scale economies, ship size economies and technical progress for the container shipping industry in Taiwan. Transp Res Part E Logistics Transp Rev 73:1–16

Xie Y, McGill MR, Dorko K, Kumer SC, Schmitt TM, Forster J, Jaeschke H (2014) Mechanisms of acetaminophen-induced cell death in primary human hepatocytes. Toxicol Appl Pharmacol 279(3):266–274

Xu D, Zhang CW, Miao Z, Cheung RK (2014) A flow allocation strategy for routing over multiple flow classes with an application to air cargo terminals. Comput Oper Res 51:1–10

Yan S, Lu CC, Hsieh J-H, Lin H-C (2015) A network flow model for the dynamic and flexible berth allocation problem. Comput Ind Eng 81:65–77

Yang C (2006) Europe and Asia standard container berth production technology application comparative analysis. China Water Transp 4(11):30–31

Yang Y-C (2014) Effect of shipping aid policies on the competitive advantage of national flagged fleets: comparison of Taiwan, Korea and Japan. Transp Policy 35:1–9

Yang ZF, Shi Y, Wang B, Yan H (2014) Website quality and profitability evaluation in ecommerce firms using two-stage DEA model. Proc Comput Sci 30:4–13

Yip TL, Liu JJ, Fu X, Feng J (2014) Modeling the effects of competition on seaport terminal awarding. Transp Policy 35:341–349

Yu QZ, Ding LY, Zhou C, Luo HB (2014) Analysis of factors influencing safety management for metro construction in China. Accid Anal Prev 68:131–138

Yusof N, Hilmy N (2014) 7.13—Radiation biology of tissue radiosterilization. In: A Brahme (ed) Comprehensive biomedical physics. Elsevier, Oxford, pp 263–287

Zavadskas EK, Turskis Z, Bagočius V (2015) Multi-criteria selection of a deep-water port in the Eastern Baltic Sea. Appl Soft Comput 26:180–192

Zhang A, Loh HS, Van Thai V (2015) Impacts of global manufacturing trends on port development: the case of Hong Kong1. Asian J Shipping Logistics 31(1):135–159

Zhang C, Xing P (2015) A research on service quality decision-making of Chinese communications industry based on quantum game. Phys A Stat Mech Appl 432(15):9–15

Zhen L (2014) Container yard template planning under uncertain maritime market. Transp Res Part E Logistics Transp Rev 69:199–217

MIX
Papier aus verantwortungsvollen Quellen
Paper from responsible sources
FSC® C105338

If you have any concerns about our products,
you can contact us on
ProductSafety@springernature.com

In case Publisher is established outside the EU,
the EU authorized representative is:
Springer Nature Customer Service Center GmbH
Europaplatz 3, 69115 Heidelberg, Germany

Printed by Libri Plureos GmbH
in Hamburg, Germany